Towards a Natural Social Contract

Patrick Huntjens

Towards a Natural Social Contract

Transformative Social-Ecological Innovation for a Sustainable, Healthy and Just Society

Foreword by René Kemp

Patrick Huntjens
Research and Innovation Centre Agri, Food
and Life Sciences (RIC-AFL)
Inholland University of Applied Sciences
Delft, The Netherlands

ISBN 978-3-030-67129-7   ISBN 978-3-030-67130-3   (eBook)
https://doi.org/10.1007/978-3-030-67130-3

© The Editor(s) (if applicable) and The Author(s) 2021, corrected publication 2021. This book is an open access publication.
Open Access This book is licensed under the terms of the Creative Commons Attribution 4.0 International License (http://creativecommons.org/licenses/by/4.0/), which permits use, sharing, adaptation, distribution and reproduction in any medium or format, as long as you give appropriate credit to the original author(s) and the source, provide a link to the Creative Commons license and indicate if changes were made.
The images or other third party material in this book are included in the book's Creative Commons license, unless indicated otherwise in a credit line to the material. If material is not included in the book's Creative Commons license and your intended use is not permitted by statutory regulation or exceeds the permitted use, you will need to obtain permission directly from the copyright holder.
The use of general descriptive names, registered names, trademarks, service marks, etc. in this publication does not imply, even in the absence of a specific statement, that such names are exempt from the relevant protective laws and regulations and therefore free for general use.
The publisher, the authors, and the editors are safe to assume that the advice and information in this book are believed to be true and accurate at the date of publication. Neither the publisher nor the authors or the editors give a warranty, expressed or implied, with respect to the material contained herein or for any errors or omissions that may have been made. The publisher remains neutral with regard to jurisdictional claims in published maps and institutional affiliations.

This Springer imprint is published by the registered company Springer Nature Switzerland AG.
The registered company address is: Gewerbestrasse 11, 6330 Cham, Switzerland

To Nicole, Talin, and Matteo.

Foreword by Prof. Dr. René Kemp

As a sustainability transition researcher, I am truly excited about this book. The book shows how the social fault lines of our times are deeply intertwined: how the social and natural world linkages raise existential concerns of security as well as justice, which call for a new social contract and transformative social-ecological innovation.

Two unique aspects of the book are that it considers bigger transformation issues (such as societies' relationship with nature, purpose, and justice) than those studied in transition studies and offers analytical frameworks and methods for taking up the challenge of achieving change on the ground. This is achieved by drawing on theories of structuration, power, governance, institutional design, and business models. The cases of nature-inclusive and regenerative agriculture, climate resilient and healthy cities, and feeding and greening megacities (in which the author is involved) are interesting cases for transition research and action research. In taking an actor-centric institutional perspective, the book addresses two mistakes: a too structuralist point of view (common in political economy) and voluntarism (common in actor-centric research of specific innovations). The author's background in conflict resolution and cooperation is a great asset. It helps to consider the political in a constructive way, through attention to justice, power, and governance.

The writing is exceptionally clear and lucid on a wide range of issues which include complex systems, reflexive and deliberative governance, transformative learning, effective cooperation, security and justice challenges, well-being, transformation literacy, and transdisciplinary research. On those issues, the writing never gets obscure or plain. This is a remarkable achievement.

With the notions of transformative socio-ecological innovation and natural social contract, the book makes an original contribution to the nature of transformative change that is needed (which goes beyond socio-technical change) and possibilities for bringing this on, through innovation, new partnerships, changes in governance, and attention to multiple value creation that jointly (in combination) make up a transition to a sustainable, healthy, and just society. If you liked the books *The Great Mindshift* of Maja Göpel and *Doughnut Economics* of Kate Raworth, you will also like this book. The same holds true if you liked the book *Transitions to Sustainable Development* by John Grin, Jan Rotmans, and Johan Schot.

Anyone interested in transformative change will find the book interesting, but I think the following readers will be particularly attracted by the book: researchers interested in doing multi-, inter-, or transdisciplinary research on transformative social-ecological innovation, reflective practitioners involved in transformative change projects, and students from universities of applied sciences who have no patience for mono-disciplinary academic research and who find the transition frameworks unduly schematic. Students of political science, political philosophy, and economics will like the discussion of transformative change (going beyond ideas and institutions) and the discussion of 'institutional design principles' for governing the commons and supporting processes of transformative socio-ecological innovation. On the last issue, the author is able to stroll further than others (Paul Mason, Paul Collier, and Mariana Mazzucato), thanks to his collaboration with Elinor Ostrom and his multidisciplinary background (which includes complex systems science, policy science, political science, biology, ecology, and environmental management).

United Nations University (UNU-MERIT)   René Kemp
Maastricht, Netherlands

Maastricht Sustainability Institute (MSI)
Maastricht University
Maastricht, The Netherlands

Acknowledgements

My research group 'Social Innovation and Governance for Sustainability' is funded by the Netherlands Ministry of Economic Affairs and Climate, as part of the 'Impact Programme: Transition in the green sector', a research programme including five research groups at four universities of applied sciences in the Netherlands.

This book is a further development of my original publication (in Dutch: *Sociale Innovatie voor een Duurzame Samenleving: Op weg naar een Natuurlijk Social Contract*), which was published in 2019 on the occasion of my inaugural address and inauguration as a Professor of Social Innovation and Governance for Sustainability, which took place on 20 June 2019 in Rotterdam, Inholland University of Applied Sciences, the Netherlands. I am grateful to Hasan Aloul (HALO Communications) and Maxime de Jong for the design and layout of the figures and tables in this book, as well as the editorial team of Springer International Publishing.

Contents

Part I   The Quest for a Natural Social Contract

1 **Introduction** 3
   1.1   Reader's Guide 7

2 **Sustainability Transition: Quest for a New Social Contract** 9
   2.1   Paradox of Prosperity 9
   2.2   Ecological Limits of Our Planet 10
   2.3   Emerging Security and Justice Challenges 14
   2.4   The Sustainability Transition: Humankind's Quest
         for a New Social Contract 19
   2.5   What's Beyond the Sustainable Development Goals? 24

3 **Towards a Natural Social Contract** 27
   3.1   What Is a Social Contract? 27
   3.2   Human Progress Without Economic Growth? 29
   3.3   Redesigning Economics Based on Ecology 34
   3.4   Debate on Role and Scope of the Free Market 37
   3.5   Anglo-Saxon Model Versus Rhineland Model 40
   3.6   Looking for a New Social Contract 43
   3.7   A Natural Social Contract 45
   3.8   Dimensions and Crossovers Within a Natural Social Contract ... 49
   3.9   TSEI-Framework for Understanding and Advancing the Process
         Towards a Natural Social Contract 67
   3.10 Development of a Natural Social Contract at Multiple
         Governance Levels 76

Part II   Theories and Concepts

4 **Conceptual Background of Transformative Social-Ecological
   Innovation** 83
   4.1   Definition of Transformative Social-Ecological Innovation
         (TSEI) 84
   4.2   Transition Studies 87
   4.3   Institutional Design Principles for Governing the Commons 90

4.4	Design Principles from Nature: Benchmarks for a Natural Social Contract	93
4.5	Complex (Adaptive) Systems	96
4.6	Adaptive, Reflexive, and Deliberative Approaches to Governance	102
4.7	Social Learning, Policy Learning, and Transformational Learning	104
4.8	Shared Value, Multiple Value Creation, and Mutual Gains	107
4.9	Effective Cooperation	110
4.10	Transdisciplinary Approach, Living Labs, and Citizen Science	111
4.11	The Art of Co-creation: Approaches, Principles, and Pitfalls	113

Part III A Research and Innovation Agenda

5 Analytical Instruments for Studying TSEI 125
   5.1   Analytical Framework for Transformative Social-Ecological Innovation (TSEI) 125
   5.2   Power and Network Analysis 130
   5.3   Framework for Analysing Different Levels of Collective Learning 133
   5.4   Collaborative Action Research 134

6 Transition to a Sustainable and Healthy Agri-Food System 139
   6.1   Challenges and Developments 139
   6.2   NWA Programme 'Transition to a Sustainable Food System' 144
   6.3   Nature-Inclusive and Regenerative Agriculture 146
   6.4   Closing the Gaps Between Citizens, Farmers, and Nature 149
   6.5   Measuring Sustainability and Health Aspects of Our Food Chains 151
   6.6   South Holland Food Family: Transition Towards a Sustainable and Self-Sufficient Food System 152

7 Governance of Urban Sustainability Transitions 159
   7.1   Urban Challenges and Developments 159
   7.2   Climate-Resilient and Healthy Cities 163
   7.3   Feeding and Greening Megacities 164
   7.4   From Linear to Circular and Regenerative Cities 165
   7.5   Collaboration for the City of the Future 167

8 Conclusion 171

Correction to: Conceptual Background of Transformative Social-Ecological Innovation C1

References 177

About the Author

Patrick Huntjens PhD. is a Professor of *'Social Innovation and Governance for Sustainability'* at the Research and Innovation Centre Agri, Food and Life Sciences (RIC-AFL), Inholland University of Applied Sciences, the Netherlands. In addition, he is Professor of 'Governance of Sustainability Transitions' at Maastricht Sustainability Institute (MSI), The School of Business and Economics (SBE), Maastricht University.

From 2017 to 2019, he was a Director of The Hague Humanitarian Cooperative for Water (HHCW), from 2013 to 2017, Head of Water Diplomacy and Climate Governance at the Hague Institute for Global Justice, and from 2011 to 2013, he was a Director of the Water Partner Foundation. At Wageningen University and Research (WUR), he was a coordinator of the Centre of Excellence—Governance of Climate Adaptation from 2010 to 2012, after working as a Coordinator of EU-Asia relations on Water Governance in the EU-funded multi-stakeholder platform ASEM Waternet (2006–2010). In the period 2000–2006, he worked as a policy officer for the Netherlands Government and as an international consultant for Royal HaskoningDHV.

With 23 years of professional experience and working in more than 40 countries, his work focuses on environmental governance and diplomacy, societal innovation, and sustainability transitions at multiple levels (global to local). He is spearheading a Research and Innovation Agenda with a core focus on the governance of sustainability transitions, with specific attention to issues of politics, power and justice in transitions, and drawing on the wider field of governance, innovation, and transition studies as well as other fields like complexity theory and systems theory. Ongoing research and educational activities include (1) transition to a sustainable and healthy agri-food system, (2) governance of urban sustainability transitions, and (3) transition to circular and regenerative economies and cultures.

Patrick has a multidisciplinary background, including a Ph.D. (Magna Cum Laude) in Complex System Sciences and Policy Sciences, an MSc degree (Cum Laude) in Political Science and International Relations, and an MSc degree in Ecology and Environmental Management. His Ph.D. dissertation was endorsed by Prof. Dr. Elinor Ostrom, the first woman in history receiving the Nobel Prize in Economics.

He solidifies his expertise with activities on the ground, mainly as international team leader, consultant, action researcher, process manager, and mediator. For example, from 2014 to 2015, he was a lead mediator (Track II) in the Israeli–Palestinian water conflict, assigned by the Geneva Initiative and the Netherlands Ministry of Foreign Affairs. Other clients include the World Bank, United Nations, European Commission, various governments, and NGOs.

Part I
The Quest for a Natural Social Contract

Introduction 1

The world moves fast. Earth's exploding global population has exasperated economic development, accompanied by wealth inequality, water and food insecurity, climate change, increased pollution, resource depletion, and loss of biodiversity as human encroachment on natural ecosystems continues. These events have all led to unparalleled economic, social, and environmental challenges with the COVID-19 pandemic as the latest deadly example. And although the pace of change may feed fear—creating a sense of powerlessness and insecurity about our shared future—these developments do not need to cause despair.

Based on scientific insights, public debate, democracy, and collective action, humankind is the only species on Earth that can deliberately change its behaviour. Our societies have enormous potential for adaptability, technological and societal innovation, and social justice. However, enacting fundamental changes will require shifting our thinking from anthropocentric social contracts and mainstream economic growth models to an ecocentric and regenerative social contract and more inclusive and deliberative approaches founded on good governance principles. This book explores these opportunities to improve the way humans live and interact with our social and natural environments.

The core philosophy of a social contract, as articulated by Aristotle, Hobbes, Rousseau, Locke, Kant, Rawls, and other political philosophers, emphasizes an implicit arrangement between citizenry, their respective societies, and legitimate government to create a healthier and safer society together. Social Contract theory states that legitimate, collective governance arrangements should be informed by the consent of the people (Weale 2004), and this theory, therefore, informs our modern concepts of democracy. The question remains, however, if current social contracts can adequately respond to the challenges of the twenty-first century. This question is more urgent when considering the current social contract focused on individualism, materialism, short-terminism, and the free market. This mindset on economic growth pays little attention to social and ecological values, as we have witnessed in the past decades. The fact that ecological vulnerability translates into social and economic vulnerability, and a complex set of security and justice challenges (Sect. 2.3), is an important omission in Social Contract theory. As Albert Einstein said: 'we cannot solve our problems with the same thinking we used when we created them'. Looking ahead, our societies will need to rethink how we inhabit and cultivate our planet and keep it healthy for future generations. Making these changes involve profound, long-term, and systemic changes in society's common practices, policies, and philosophies that will rely on new knowledge and skills.

The nature of the social, environmental, and economic problems we face today requires a new social contract, a Natural Social Contract. A Natural Social Contract does justice to a human being's natural state (human life is group life) and to the natural position of humankind and society within a larger ecosystem, that of planet Earth. The Natural Social Contract regards society as a social-ecological system, focusing on people as members of a community and as part of a natural ecosystem. It emphasizes long-term sustainability and general welfare by combining human and nature, and recalibrating our unfettered approach to unlimited economic growth, overconsumption, and over-individualization. The end result, I argue, is for the benefit of ourselves, our planet, and future generations.

'Towards a Natural Social Contract' poses several thought-provoking questions about human nature, our relation to social and natural environments, and how we humans have shaped and organized our societies.

How would Mother Nature judge humankind? Would she be proud or concerned? Would she agree with Friedrich Nietzsche saying, 'Our planet is sick, and the disease is called Man'? Or would she view us as children or adolescents who seek thrills and take risks? They have to, she might say, because they learn from it. But perhaps Mother Earth thinks it's time for us to mature, clean up our mess, and take responsibility.

This brings me to a fundamental question of Political Philosophy. Is current society a reflection of true human nature, or did we somehow along the way lose sight of our true nature? Is current society really the best we can think of? In this book I argue that the divide between humans and nature that arose during the Enlightenment, and the capitalist economic logic and related economic structures that were put in place after the Second World War, have blurred or ignored several important core values. These include social and environmental stewardship, planetary health, environmental security and justice, intergenerational justice and equity, and the Rights of Nature. Hence, do we prefer to consider ourselves a 'Homo Economicus', namely a species that places more value on individualism, self-interest, material wealth, privatization, short-term gains, and a free-market economy focused on profit and economic growth that erodes social and ecological values? Or do we prefer to consider ourselves as a 'Homo Ecologicus'? A species that puts more value on unity, solidarity and connectivity, sustainable co-management of the Commons, social and environmental stewardship, human security, planetary health, environmental protection, and achieving justice, human rights, and the Rights of Nature?

I argue for an approach that draws out the best in people and our societies. An approach that facilitates a transition from ego-awareness to eco-awareness and considers humans as a 'Homo Ecologicus' rather than 'Homo Economicus'. This approach will help us restore our balance with our own nature and with planet Earth. An approach where Nature serves as our guide, teacher, companion, and inspiration, and not as our enemy or obstacle to be dominated or controlled by humans to serve the exclusive needs of humanity.

A Natural Social Contract as proposed in this book (Sects. 3.7 and 3.8) is an open and broad theoretical framework across multiple dimensions (i.e. social, ecological, economic, and institutional), which serves to start a dialogue about ways to improve the current social contract, targeting a more sustainable, regenerative, healthy and just society. It can help policymakers, administrators, and decision makers, concerned citizens and professionals to make better decisions about how to organize our twenty-first-century society.

This book explains how Transformative Social-Ecological Innovation (TSEI) plays a central role in the sustainability transition and humankind's search for a Natural Social Contract. Transformative Social-Ecological Innovation is defined as 'systemic changes in established patterns of action and in structure, including formal and informal institutions and economies, that contribute to sustainability, health and justice in all social-ecological systems' (definition by author). Creating a sustainable and healthy future for societies will require institutional change as well as multiple parties, multiple sectors, and multiple levels of government to act and collaborate effectively. TSEI is based on processes of collective learning and co-creation in which different but interdependent parties learn to develop new knowledge and solutions in a transdisciplinary approach.

From an economic perspective, the most fundamental systemic change required for realizing a Natural Social Contract is a transition from our current linear economic system (i.e. produce, use and dispose) towards circular and regenerative economies and cultures. The promise of a circular and regenerative economy is to organise sustainability, circularity and social justice at different scales, preferably as an integrated economic and social endeavour, which involves technological, social, organisational and institutional innovation. In practice, this will require a radical change from linear to circular business models characterized by collective and shared value creation. Innovative and hybrid forms of financing, such as revolving energy and sustainability funds, will also be a part of this development. Likewise, the joint management of commons (instead of private ownership) and a sharing economy improving access to goods and services would offer important systemic changes toward a Natural Social Contract and in turn boost efficiency, sustainability, and community values.

In Part 2 of this book, I introduce and define the concept of Transformative Social-Ecological Innovation (TSEI) (Sect. 4.1) and apply its use in the complex multi-actor and multi-level context of the sustainability transition. Based on a literature review, I have highlighted key theories and concepts that add substance to the workings of TSEI. This includes transition studies (Sect. 4.2), institutional change and the structure-agency debate (Sect. 3.9), resilience theory and social-ecological systems (Sect. 3.8), institutional design principles for governing the commons (Sect. 4.3), design principles from nature (Sect. 4.4), complex adaptive systems (Sect. 4.5), adaptive, reflexive, and deliberative approaches to governance, management, and planning (Sect. 4.6), social learning, policy learning, and transformational learning (Sect. 4.7), shared value, multiple value creation, and mutual gains approach (Sect. 4.8), effective cooperation (Sect. 4.9), quintuple helix innovation model (Sect. 3.9), transdisciplinary cooperation, living labs, and citizen science (Sect. 4.10), and finally, a section on the art of co-creation: approaches, principles, and pitfalls (Sect. 4.11).

Drawing on the insights from this literature, I argue that studying Transformative Social-Ecological Innovation should involve both structure and agency, in particular a focus at decisive moments where both structure and agency intersect (i.e. in action situations). This also includes outputs, outcomes, and impacts. I identify a critical need to focus on the fundamentally political character of TSEI and the need for multiple value creation for parties to identify shared values, mutual gains, and common interest.

These findings from literature have been brought together in a conceptual framework (Sect. 3.9) and an analytical framework (Sect. 5.1) for Transformative Social-Ecological Innovation (TSEI). The TSEI-framework is proposed as an open framework. In that sense, TSEI accounts for additional predictors and moderators if they have a documented effect. The framework can also be used for institutional and political-economic analyses, with a special focus on the power dynamics at play (Sect. 5.2). Power dynamics can be studied by looking at series or clusters of closely related action situations in which the initiation, format, content, and output of each action situation are analysed. To further support the practical applicability of the TSEI-framework, an analytical framework for different levels of collective learning has been operationalized (Sect. 5.3).

In Part 3, I present a Research and Innovation Agenda with various analytical instruments (Chap. 5) and an overview of relevant and ongoing research and educational activities, including Transition to a sustainable and healthy agri-food system (Chap. 6), and Governance of urban sustainability transitions (Chap. 7).

The Transformative Social-Ecological Innovation (TSEI) framework offers new ideas for unpacking and understanding institutional change across sectors and disciplines and at different levels of governance. To this end, it identifies intervention and leverage points and helps to formulate sustainable solutions that can include different perspectives, as well as changing and competing needs. Overall, a new Natural Social Contract and the concept of TSEI encourage public officials, business leaders, and the greater public to consider how society can concretely improve humankind's response to our greatest challenges.

If you are concerned about our society and our planet, and keeping both healthy for future generations, then this book is written for you. And if you have an interest in the systemic changes required to fundamentally shift our social, economic, ecological, and institutional perspectives, this book is for you too. Together, we can promote a sustainable, healthy, and just society and achieve change on the ground. This book offers a way forward.

1.1 Reader's Guide

This book is intended for academics and broader audiences alike. Policymakers, civic leaders, entrepreneurs, and the public will find practical insights and philosophies along with more in-depth theoretical discussions summarized in outline.

The book will also appeal most to individuals engaged in multi-, inter-, and transdisciplinary research on Transformative Social-Ecological Innovation, and reflective practitioners involved in transformative change projects. A wide readership of students, researchers, and policymakers interested in social innovation, transition studies, social policy, development studies, social justice, climate change, environmental studies, political science, and economics will find this cutting-edge book particularly useful.

In Chap. 2, I provide a problem definition and the related field of development. I will start with an introduction to the paradox of prosperity (Sect. 2.1), the ecological limits of our planet (Sect. 2.2), and how this relates to a broad range of security and justice issues (Sect. 2.3). Following this, the chapter addresses the necessity and nature of the sustainability transition (Sect. 2.4). Chapter 2 concludes with a plea to be more explicit on the future beyond the sustainability transition (Sect. 2.5).

In Chap. 3, I explain how the sustainability transition offers humankind an opportunity for a new social contract: a 'natural' social contract. Following a brief introduction on the origins of the social contract (Sect. 3.1), I address the question of whether there can be human progress without economic growth, and explore redesigning economics based on ecology. This chapter includes a debate on the role and scope of the free market (Sect. 3.4), as well as an examination of how the Anglo-Saxon and Rhineland models fare in this debate (Sect. 3.5). Chapter 3 will also describe why we need a new social contract and what it should entail (Sect. 3.6). In doing so, I will embark on a quest for a Natural Social Contract (Sect. 3.7), and I will describe its theoretical foundations with multiple dimensions and crossovers (Sect. 3.8). In order to gain a better understanding of a Natural Social Contract and boost the development of such an arrangement, this chapter presents a conceptual framework for Transformative Social-Ecological Innovation (TSEI) (Sect. 3.9), and how this may transpire at various governance levels (Sect. 3.10).

Part 2 of the book provides a brief literature review on the conceptual background of Transformative Social-Ecological Innovation (Chap. 4). This includes a survey of key theories and concepts such as transition studies, institutional design principles for governing the commons, design principles from nature, various approaches to collective learning, multiple value creation, effective cooperation, and a section on the art of co-creation among others.

Part 3 offers a research and innovation agenda for a better understanding and advancement of Transformative Social-Ecological Innovation towards a sustainable, healthy, and just society. Chapter 5 highlights several analytical instruments for studying Transformative Social-Ecological Innovation, including an analytical framework for Transformative Social-Ecological Innovation (Sect. 5.1), a power and network analysis (Sect. 5.2), a framework for analysing different levels of collective learning (Sect. 5.3), and a section on collaborative action research (Sect. 5.4).

Chapters 6 and 7 will underscore relevant and ongoing research and educational activities, including the transition to a sustainable and healthy agri-food system (Chap. 6) and urban sustainability transitions (Chap. 7).

Finally, Chap. 8 wraps up the book with a conclusion, followed by a bibliography.

Open Access This chapter is licensed under the terms of the Creative Commons Attribution 4.0 International License (http://creativecommons.org/licenses/by/4.0/), which permits use, sharing, adaptation, distribution and reproduction in any medium or format, as long as you give appropriate credit to the original author(s) and the source, provide a link to the Creative Commons license and indicate if changes were made.

The images or other third party material in this chapter are included in the chapter's Creative Commons license, unless indicated otherwise in a credit line to the material. If material is not included in the chapter's Creative Commons license and your intended use is not permitted by statutory regulation or exceeds the permitted use, you will need to obtain permission directly from the copyright holder.

Sustainability Transition: Quest for a New Social Contract

This chapter will provide an overview of the necessity and nature of the sustainability transition, starting with the paradox of prosperity (Sect. 2.1), the ecological boundaries of our planet (Sect. 2.2) and how this relates to a broad range of security and justice issues (Sect. 2.3). Following this, the chapter provides a brief description of the nature of the sustainability transition (Sect. 2.4), and concludes with an argumentation to be more explicit on what comes after the Sustainable Development Goals (SDGs) of the UN 2030 Agenda (Sect. 2.5).

2.1 Paradox of Prosperity

Economies around the world are usually designed for one purpose: economic growth. In recent decades, the free market has flourished, and though it has brought tremendous economic prosperity to society in the process, it also has major downsides. The positive prospects for globalization and economic growth that spurred people on in the 1990s have made way for uncertainty, an actual crisis (the 2008 global credit crisis), and fears about the future. Already in 2006 the Stern Review on the Economics of Climate Change concluded that: 'Our actions over the coming few decades could create risks of major disruption to economic and social activity, later in this century and in the next, on a scale similar to those associated with the great wars and economic depression of the first half of the twentieth century' (Stern Review 2006, page xv).

Never before has humankind been confronted with the negative consequences of its own actions on such a large scale. Growing wealth inequality, depletion of natural resources, pollution of water, land, and air, climate change, loss of biodiversity, malnutrition, and (often within one country) diseases of affluence such as obesity and diabetes type II, financial crises (in 2008 and 2020), epidemics and pandemics (including Avian Flu, SARS, MERS, Corona-virus), trade wars (e.g. between the USA and China), and migration challenges (e.g. Syria, and climate change-related refugees in many parts of the world) are but some of the issues we face today. We are now discovering that the ecological vulnerability translates into social and economic vulnerability. These problems feed fear, powerlessness, and uncertainty about developments that individuals do not seem to be able to control. Accordingly, the downside of prosperity has major consequences for society and the planet. This is also known as the paradox of prosperity. I will follow-up on the economic debate behind this paradox in Chap. 3.

By way of illustration, recent research has shown the increase in economic capital in the Netherlands has been out of step with the country's 'broad prosperity' since the 1970s (Lintsen et al. 2018). Broad prosperity includes the economic, ecological, and social aspects of prosperity, such as education, health, good governance, social equality, the quality of the environment, and natural capital. The Broad Prosperity Monitor (CBS 2020) paints a troubling picture of the trend of broad prosperity in the Netherlands (see Fig. 2.1), with indicators related to natural capital steadily declining or not improving. Also trends in human and social capital are out of step with the increase in economic capital. The report shows that using economic growth as a compass for government policy could, in the long run, have disastrous consequences (Lintsen et al. 2018).

The Broad Prosperity Monitor also looks at broad prosperity elsewhere, i.e. the effects that Dutch society has on the rest of the world. Again, the results are worrying. The trends in this area show that the Netherlands has started using more and more fossil fuels and biomass originating from the rest of the world, the least developed countries in particular (CBS 2018, 2020). In many African countries in particular, this trade in natural resources leads to problems that can reduce broad prosperity in those countries, and often mainly benefit a small elite; a phenomenon called the 'resource curse' in literature (ibid).

2.2 Ecological Limits of Our Planet

The necessity of a Natural Social Contract underlying a sustainable society becomes clear when we look at the ecological boundaries of our planet.

The planet's ability to sustain humankind is put under increasing pressure, primarily due to the growing world population, economic growth, large-scale pollution, depletion of natural resources, and climate change. In order to keep our planet healthy for future generations, we will have to accurately map out and respect our planet's boundaries. This is by no means an easy task. Johan Rockström et al. (2009) have mapped nine of our planet's boundaries: climate change, loss of biodiversity, excess nitrogen and phosphorus production, stratospheric ozone deposition, ocean acidification, global freshwater consumption, changes in land use caused by agriculture, air pollution, and chemical pollution.

While three of those boundaries had already been exceeded in 2009, a follow-up study published in Science in 2015 claimed that 4 of the 9 planetary boundaries have already been exceeded as a result of human activity, namely climate change, loss of biosphere integrity (in 2009 called: loss of biodiversity), changes in land use caused

2.2 Ecological Limits of Our Planet 11

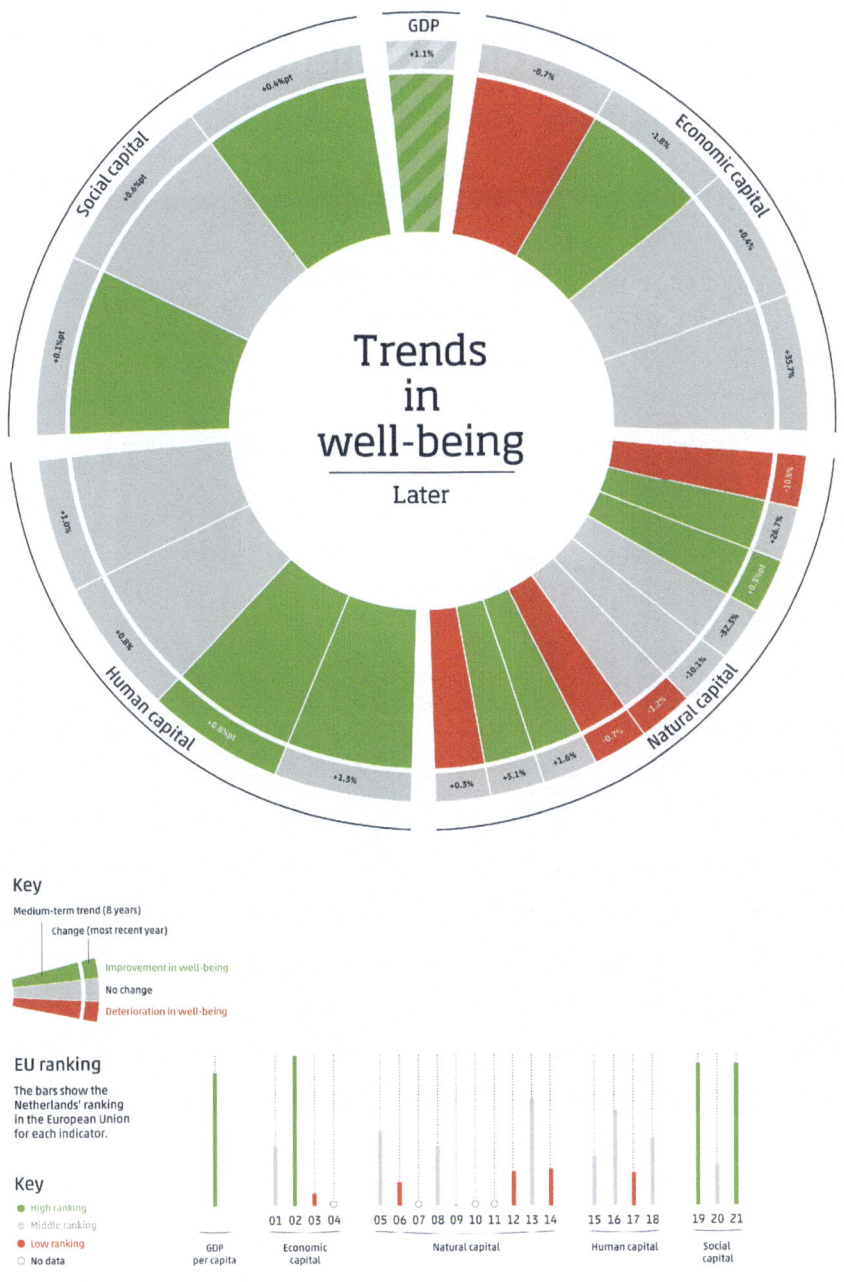

Fig. 2.1 Broad Well-being Trends (Central Bureau of Statistics 2020)

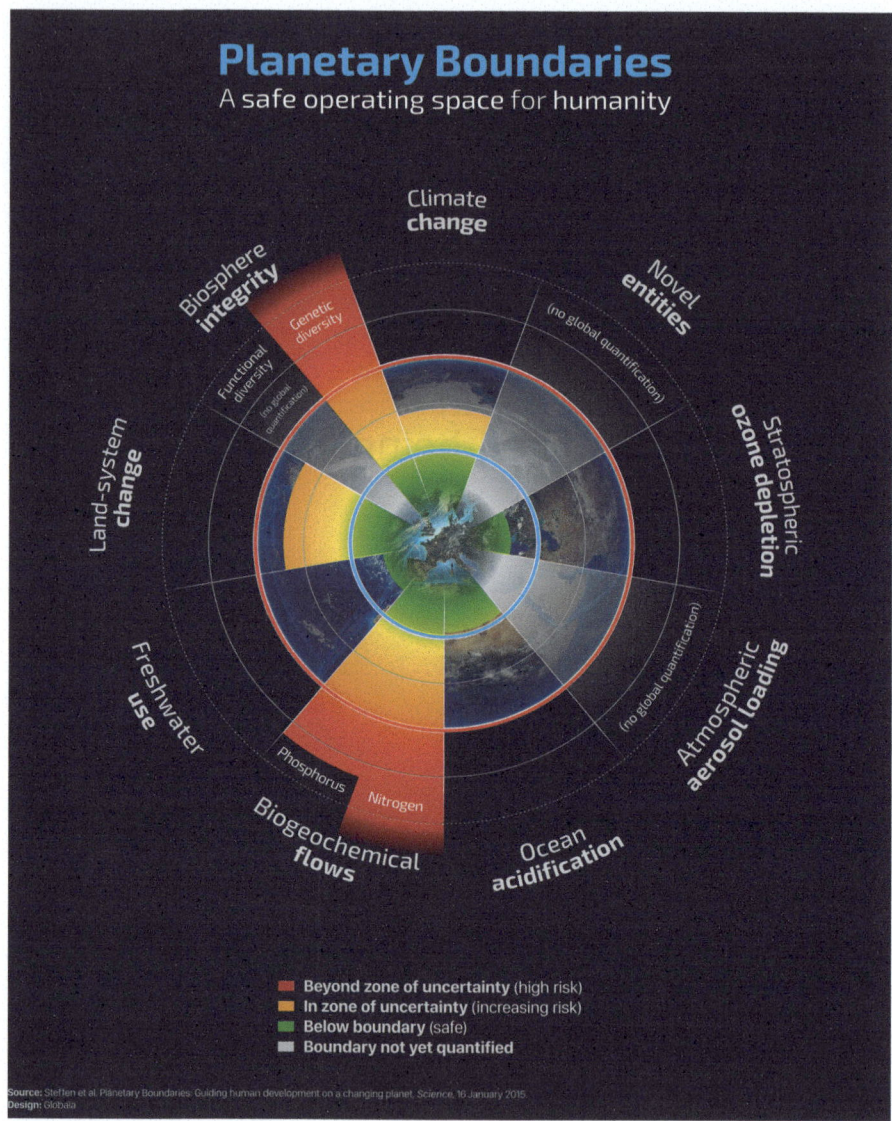

Fig. 2.2 Boundaries for nine planetary systems (Steffen et al. 2015). The wedges represent an estimate of the current status of each variable

by agriculture, and excess nitrogen and phosphorus production (Steffen et al. 2015). See Fig. 2.2.

Humankind has triggered a biodiversity crisis that is no less severe than the climate crisis. A report of the UN biodiversity panel (IPBES 2019) shows that without rapid, far-reaching measures, hundreds of thousands of plant and animal species will become extinct and vital ecosystems will decline. 'The liveability of our planet is at risk', so concludes the UN report on biodiversity (IPBES 2019). Loss of biodiversity is at least as great a threat to humankind as global warming, but the UN is concerned that the urgency of halting declining biodiversity and ecosystems is much less keenly felt. This, however, is at our own peril, the UN report concludes, because the decline is a direct consequence of our consumption patterns and nature and biodiversity are essential for food production, our water supply, medicine supply, and general public safety and social cohesion (ibid.).

The 1800-page UN report paints a gloomy picture of the many ways in which humankind is plundering nature and undermining its ecosystem services (the benefits that living nature confers on humankind). More than 75% of all land, 40% of the oceans, and 50% of our rivers have been degraded due to deforestation for agricultural and livestock farming purposes, by mining, urbanization, infrastructure, and fishing. Only 13% of all land and 23% of the oceans are still more or less untouched. More than 20% of all agricultural land is degraded (IPBES 2019).

Ecosystem pollution also takes its toll. For example, more than 80% of the world's wastewater is not treated. In addition, an estimated 300 to 400 million tonnes of heavy metals and other toxic substances are dumped into the environment, as a result of which 40% of the world's population does not have access to clean drinking water, to name but one consequence. Millions of tonnes of plastic disappear into the oceans every year (ibid.).

Nature and biodiversity are essential in our fight against climate change. Forests and oceans absorb half of our carbon emissions. Over the past 5 years, however, deforestation has wiped out rainforest equivalent to five times the size of England for agriculture in order to meet the world's needs for beef, soybeans, palm oil, and biofuels (ibid.).

'Climate change has been called the single biggest challenge for humanity over the coming centuries (UNSG 2014, 2018; McKibben 2019). Given the scale of the problem, its impacts on human life, and the level of coordinated action required to address it, this statement seems more than justified. After the Intergovernmental Panel for Climate Change (IPCC) published its first assessment report in 1990 it was accused of dramatizing the anthropogenic causes as well as the potential effects of global warming; now we know that the researchers had in fact underestimated both causes and effects. Although uncertainty and unpredictability remain, the scientific basis of climate change is now well established. It suggests that change is happening more quickly than previously estimated and can no longer be framed as a distant threat (Stern 2013). The past three decades have likely been the warmest 30 years of the last 1400 years (IPCC 2013). The atmospheric concentration of greenhouse gases (GHG) has increased to a level unprecedented in the last 800,000 years, and their "mean rates of increase" over the past century "are, with very high confidence, unprecedented in the last 22,000 years" (*idem*). Changing precipitation patterns, melting ice caps, rising sea levels, acidification of oceans, and heightened climatic variability are only some of the predictable consequences of a climate destabilized by warming atmosphere and oceans' (cf. Huntjens and Nachbar 2015; Huntjens et al. 2018).

In October 2018, the United Nations Intergovernmental Panel on Climate Change (IPCC) published a landmark report that concluded that governments must take urgent, unprecedented, and far-reaching action by 2030 in order to limit global warming to a maximum of 1.5 degrees Celsius. However, the target set in the Paris climate agreement of a maximum of 1.5 degrees of warming will almost certainly not be achieved, such is the painful conclusion of the UN climate panel (IPCC 2018).

Recent information from the World Meteorological Organization, the World Bank, and the International Energy Agency shows the relentless pace of climate change (WMO 2019).

With global warming, we are now seeing deadly heat waves and massive wildfires in some parts of the world, while the largest physical structures on our planet, such as coral reefs, ice caps, and rainforests, disappear before our eyes (McKibben 2019). António Guterres, the United Nations secretary general, said, 'I am beginning to wonder how many more alarm bells must go off before the world rises to the challenge' (UNSG 2018).

The UN climate panel notes that global warming is currently more likely to reach 3 degrees than 2, let alone the targeted 1.5 degrees. This is because the 195 countries that signed the Paris Climate Agreement have not yet devoted enough effort to reducing their greenhouse gas emissions. As of 2018, global carbon emissions amount to some 42 billion tonnes. If we keep going at the current rate, we will have reached 1.5 degrees of warming in about 15 years from now (IPCC 2018).

Based on the above, we can undoubtedly state that humankind has wreaked havoc on the planet. 'Our planet is ill, and the disease is called man'. This is the diagnosis made by Friedrich Nietzsche, one of the brightest philosophers in European history, some 150 years ago. As a species, humans can be compared to a parasitic infectious disease that kills its own host, i.e. our planet. This image stands in stark contrast to the common belief that humankind is the highest, most developed species in nature. The great existential question of our time is whether humankind can shift from being a parasitoid to a symbiotic species that can live sustainably with its host, i.e. planet earth, on time. The answer to that question is an emphatic YES, provided that people as individuals, groups, and societies, from global to local, are willing to immediately and effectively make work of the sustainability transition.

There is every reason to unite our efforts and work on this issue together. Humankind is the only species on this earth that—based on scientific insights, public debate, and collective action—can deliberately steer its behaviour. This means society has enormous potential for technological and societal innovation and adaptability.

2.3 Emerging Security and Justice Challenges

Security and justice mean different things to different groups and individuals and the potential implications of climate change, resource depletion, and environmental degradation for security and justice are varied and complex (Huntjens et al. 2018).

2.3 Emerging Security and Justice Challenges

Both the security and justice implications of climatic change and environmental degradation have been subject to an increasingly broad debate in the scientific as well as the policy community. Despite this increase in attention, the ways in which the effects of the ecological crisis will impact security and justice at various levels are still far from clear (*ibid.*).

'Human security as a concept aims to capture the broad range of factors that determine people's livelihoods and their ability to exercise their human rights and fulfil their potential. The UNDP's 1994 Human Development Report definition argues that the scope of global security should be expanded to include threats in seven areas: economic, food, health, environmental, personal, community and political security (UNDP 2011). For instance, climate change is understood as a threat to human security in that it disrupts individuals' and communities' capacity to adapt to changing conditions, usually by multiplying existing or creating new strains on human livelihoods (Brauch and Scheffran 2012; Barnett and Adger 2007). The various effects of global warming, resource depletion, and environmental degradation are already being felt as real consequences for real people and communities around the world. Because it changes ecosystems that form the basis not just for plant and animal but human life, climate change is a development that goes to the very heart of human coexistence and confronts us with challenges concerning the security as well as justice of our societies' (cf. Huntjens and Nachbar 2015; Huntjens et al. 2018).

The human security approach emphasizes 'the interconnectedness of both insecurities and responses. Insecurities are interlinked in a domino effect in the sense that each insecurity feeds on the other. If not managed proactively, these can spread to other regions or countries. For example, climate change may induce drought, giving rise to food insecurity with impacts on health, while competition over scarce resources threatens community cohesion, and personal and political security' (cf. United Nations 2016). Besides human security, other security dimensions such as planetary security, as well as national security and international security (the security of states) need to be taken into account. An overview of security challenges is provided in Table 2.1.

'A broad range of human rights is affected by climate change and environmental degradation, including the rights to life, freedom of movement, housing, water, food, health or professional development. Consequently, they ought to be dealt with in national and international human rights bodies, such as the European Court of Human Rights, the European Parliament subcommittee on human rights, the Council of Ministries at the Council of Europe and during UN Human Rights Council Sessions. A human rights-centred approach shifts the focus from purely economic and scientific considerations and consequences towards human rights violations caused or exacerbated by climate change or environmental degradation. This approach enhances democratization through active citizen participation and the claim for transparency and accountability. Thus, a positive side-effect of such responses to climate change is the creation of new ways of governance seeking justice based on good governance principles' (cf. Huntjens et al. 2018).

Table 2.1 Security challenges (non-exhaustive) related to the various effects of global warming, resource depletion, and environmental degradation

Security challenge	Brief explanation
Planetary security	Four planetary boundaries have already been exceeded as a result of human activity, namely climate change, loss of biodiversity, changes in land use caused by agriculture and excess nitrogen and phosphorus production (Steffen et al. 2015).
Food security	Over the coming decades, a changing climate, growing global population, rising food prices, and environmental stressors will have significant yet uncertain impacts on food security. Currently, about 2 billion people in the world experience moderate or severe food insecurity (FAO, IFAD, UNICEF, WFP, and WHO, 2019). The lack of regular access to nutritious and sufficient food that these people experience puts them at greater risk of malnutrition and poor health. Although primarily concentrated in low- and middle-income countries, moderate or severe food insecurity also affects 8 percent of the population in Northern America and Europe (ibid.).
Water security	*'There are many factors affecting water security, including a growing population, agricultural irrigation, rising domestic demand due to rising standard of living, increasing industrial demand, escalating energy consumption, mining, climate change, urbanization, deforestation, and migration of people'* (cf. Singh 2017). Water Security is defined by UN-Water (2013) as 'The capacity of a population to safeguard sustainable access to adequate quantities of acceptable quality water for sustaining livelihoods, human well-being, and socio-economic development, for ensuring protection against water-borne pollution and water-related disasters, and for preserving ecosystems in a climate of peace and political stability'. *For more on water security see Pahl-Wostl et al. (2016)*
Energy security	Climate change tends to negatively affect the power sector, inter alia, by causing cooling problems in power plants and impairing the water supply required for hydropower generation (Van Vliet et al. 2013; Rübbelke and Vögele 2013).
Economic security	According to the Global Risk Report (World Economic Forum 2020) the top five of risks to global economy (in terms of likelihood) are: (1) Extreme weather, (2) Climate action failure, (3) Natural disasters, (4) Biodiversity loss, and (5) Human-made environmental disasters. From a business perspective, climate change poses wide-ranging threats to business operations, such as reduction/disruption in production capacity and supply chains, increased operational costs, or inability to do business, with the latter usually resulting in loss of jobs.
Environmental security	'Environmental security is the proactive minimization of anthropogenic threats to the functional integrity of the biosphere and thus to its interdependent human component' (cf. Barnett 1997). There multiple threats to environmental security, such as resource scarcity (diminishing supplies of inputs into human systems) and pollution (the contamination of inputs into human systems), occurring at multiple scales (from global to local). (Barnett 2009). 'The condition of environmental security is one

(continued)

2.3 Emerging Security and Justice Challenges

Table 2.1 (continued)

Security challenge	Brief explanation
	in which social systems interact with ecological systems in sustainable ways, all individuals have fair and reasonable access to environmental goods, and mechanisms exist to address environmental crises and conflicts' (cf. Glenn et al. 1998).
Health security	Environmental degradation can have a significant impact on human health. 'Air pollution and exposure to hazardous chemicals are important causes of the environment-related burden of disease in OECD countries' (cf. OECD 2011). 'The transport and energy sectors are major contributors to air pollution, while important sources of chemical pollution are agriculture, industry, and waste disposal and incineration' (cf. ibid.). Furthermore, health security is threatened by environmental stressors such as malnutrition and insufficient access to health services, clean water, and other basic necessities.
Community, personal and political security	'Competition over scarce resources threatens community cohesion, and personal and political security (United Nations 2016). The effects of climate change, particularly climate change related environmental impacts and associated resource scarcity, and following migration of people once coupled with other structural and socio-political factors can contribute to exacerbate existing conflicting relations between parties in destination area' (cf. Islam and Nur 2019). In addition, 'rapid onset events (such as storms, floods, and bush fires) and slow onset events (such as droughts, water scarcity, sea level rise, desertification, and coastal erosion) place stress on those who are already vulnerable, such as indigenous peoples, women, and children. These groups may be more dependent on natural resources and a healthy ecosystem for their survival and, in addition, may have less access to coping mechanisms (e.g. mobility, land ownership, and emergency funds) in their place of residence. As a result, they become refugees, migrants, or Internally Displaced Persons (IDPs) as an adaptation strategy' (cf. Huntjens et al. 2018). Since 2008, an average of nearly twenty-seven million people have been displaced annually by natural hazard-related disasters. This is the equivalent to one person being displaced every second (IDMC 2015). The effects of climate change are expected to intensify such disasters and accelerate displacement rates in upcoming decades (ibid.).
National and international security	Climate change is a threat to national and international security (the security of states) in two ways. Firstly, climate change contributes to higher instability in some of the world's most volatile regions. Secondly, climate change can contribute to tensions in stable regions (CNA Military Advisory Board 2007; Huntjens et al. 2018). Most strategic documents that have established a link between climate change and state security have also emphasized the need for disaster preparedness and measures aimed at building resilience in countries at risk of climate-induced conflict (Brzoska 2012)

'A range of immediate and effective mechanisms is needed to safeguard the rights of future generations and protect them from the potential negative implications and harm caused by climate change and environmental degradation. Polly Higgins and colleagues argue that the rights of nature itself must be protected against ecocide (Higgins 2010; Gauger et al. 2012; Higgins et al. 2013; Lay et al. 2015), and propose to criminalize human activities that cause extensive damage to ecosystems. One proposed step forward could be the legal acknowledgement of the crime of ecocide as the fifth Crime Against Peace, the other four being genocide, crimes against humanity, war crimes, and the crime of aggression as set out in the Rome Statute of the International Criminal Court. The United Nations has discussed the crime of ecocide for decades (see Gauger et al. 2012), although the chances of succeeding in modifying the Rome Statue are small, particularly as countries with large fossil fuel interests or with large fossil fuel related multinationals are likely to vote against, afraid of being held accountable and combined with a complex set of legal complications' (cf. Huntjens and Zhang 2016).

In contrast to international law, a more effective approach, so far, is illustrated 'by a variety of lawsuits on climate change and environmental degradation in national jurisdictions, while several states have started to recognize the rights of nature, ecosystems, and animals and there has been an increasing recognition of the intersection between human rights and environmental degradation' (cf. Huntjens et al. 2018). Significantly, several court judgements afford protection to ecosystems and animals (ibid.), while several countries have recognized Rights of Nature in their legal frameworks and/or jurisprudence, e.g. in Uganda, Peru, Ecuador, Mexico, Colombia, India, Bangladesh, New Zealand, and communities in the USA.

More recently, lawsuits to force countries towards an effective climate policy are increasingly being considered as an important avenue for breaking through political indifference and deadlock (ibid.). The verdict in the court case on climate justice in the Netherlands is the first of its kind worldwide. When filing the court case, civil society organization Urgenda argued that the government is doing too little and should be held accountable for not taking appropriate action to safeguard a healthy environment for future generations. In particular, Urgenda claimed that the Netherlands must reduce greenhouse gases drastically by 2020, and much more than agreed to within the EU. In its defence the State of the Netherlands made an appeal to EU policies and international agreements. In addition, the State relied on the separation of powers: it claimed that political decisions on climate policy should not be taken in court, but by the government and parliament. The judge argued that independent courts sometimes need to decide on the conduct of politics, but it must be done with reticence. Since mitigation of climate change actually requires a reduction of 25 to 40 percent of GHG in 2020, with explicit reference to scientific consensus on this topic, the judge found 25 percent a modest and thus reticent requirement, when compared to the upper target of 40 percent. As a result, the State of the Netherlands was legally forced to reduce greenhouse gas emissions with at least 25% by 2020, much more than its own government plans, which was about 17% by the time of filing the court case. Never before has a court sentenced a national government to a more effective climate policy.

While it is important to recognize the progress made by national jurisdictions in addressing crimes to the environment, a void within international law remains and affects the ability of domestic jurisdictions to respond to grave problems of climate change. For instance, 'the Arctic is still protected by soft law instruments only, and the legal regime protecting the environment against reckless exploitation remains inadequate' (cf. Lay et al. 2015).

2.4 The Sustainability Transition: Humankind's Quest for a New Social Contract

The five decades from 2000 to 2050 will go down in history as the sustainability transition. The sustainability transition constitutes a search for a new social contract, in this book coined as a Natural Social Contract (see Chap. 3). The core philosophy behind a social contract is that the members of a society enter into an implicit contract with the goal of living a better, safe life together (Kalshoven and Zonderland 2017). Such a contract includes agreements about public goods and services, for instance, as well as taxes, detailing how everyone contributes to and benefits from society. The purpose of the social contract is serving the common or greater good to ensure the sustainability of the society in question and protect the individuals within it. In other words, the social contract is expected to provide security and justice for all (see Sect. 3.1 for more details).

The global and systematic nature of the environmental problems we face today necessitates fundamental changes in key societal, economic, and legal systems. Making these changes, however, will require much more than step-by-step efficiency improvements. Rather, we will have to realize profound, long-term changes in dominant practices, policies, and philosophies that, in turn, will require new knowledge.

Transition is defined as a fundamental change in the structure (institutional, physical, and economic structure), culture (shared ideas, values, and paradigms), and methods (routines, rules, behaviour) of a system (Rotmans 2005). The sustainability transition will have to include changes in all these dimensions if we are to leave the earth in better shape for future generations. This also means overcoming short-termism and a singular focus on economic growth, which currently dominate political and economic thought. On the flip side, the list with counter-proposals to unlimited economic growth has grown rapidly and the rollout is getting stronger, in particular since the 2008 global credit crisis (see Sect. 3.2). The Corona-crisis is expected to become the next major tipping point towards a sustainable society, illustrated by an observation from EU-officials that the new EU Green Deal is expected to have an accelerated implementation due to the devastating impact of the Corona-crisis on industries relying on burning fossil fuels, such as the car and aviation industry, transportation, agriculture, construction, and electricity production. In any case, the granting of substantial EU aid packages to keep the economy going during the Corona-crisis will have to go hand in hand with hard conditions for a transformation towards a green economy, according to the EU.

The call for fundamental societal changes targeting sustainability is high on local, national, and international agendas. At the global level, the United Nations have set Sustainable Development Goals, as part of its 2030 Agenda, comprising both socio-economic and environmental dimensions of sustainability. In Europe, 'a good and healthy life in 2050 within our planet's ecological boundary' is a core component of environmental policy (EU, 7th Environment Action Programme 2013). This vision has also been incorporated in other lines of EU policy.

In the past two decades, the European Union has introduced a large body of environmental legislation, which has succeeded in significantly reducing air, water, and soil pollution. Legislation on chemicals has been modernized and the use of many toxic or hazardous substances has been restricted. Today, EU citizens boast the highest-quality water in the world, and more than 18% of the land mass covered by the EU has been designated as a protected area. However, many problems remain and must be tackled in a structured way (EU, 7th Environment Action Programme 2013). To solve these problems and achieve the goals set in environmental policy, the EU will need to make far-reaching changes in its production and consumption systems. The Low-Carbon Economy Roadmap, for instance, aims to reduce greenhouse gas emissions in the EU by 80% by 2050, while the Circular Economy Strategy focuses on significant improvements in waste reduction and management by 2030. In December 2019, 'the European Commission released the European Green Deal, a blockbuster policy aimed at halting climate change by shifting to clean energy and a circular economy, thereby increasing resource efficiency and restoring biodiversity. The agreement will establish a €100bn "Just Transition Mechanism" and urge European countries to set up a broad national tax reform Mechanism, with "climate taxes" as the focus' (cf. UN-Habitat 2020).

Within the sustainability transition, we can identify three important systemic changes:

1. **climate and energy**: greenhouse gas emissions (such as carbon dioxide, methane, and nitrous oxide) must be reduced drastically. Fossil energy must be replaced en masse by clean energy. Climate change mitigation and adaptation is required in almost every sector.
2. **agriculture and food**: the quality of nature, water, and air must be improved without compromising the production of a sufficient supply of healthy, sustainable, and safe food. All around the world, we will have to feed approximately 10 billion mouths by 2050.
3. **circular and regenerative economies and cultures**: the depletion of raw materials and the continued undermining of ecosystems must be prevented.

This transition to a new, sustainable economy and society has many faces. The next economy will be a digital, bio-based, circular, sharing, maker and robot economy (RNE 2016). The new economy will be increasingly based on horizontal relationships and small-scale, locally organized networks of producers and consumers rather than vertically integrated structures (ibid.), but the new economy will also be characterized by great uncertainty and disruptive developments (ibid.).

2.4 The Sustainability Transition: Humankind's Quest for a New Social Contract

Many traditional sectors, such as the fossil fuel industry and companies that fail to make serious work of sustainability, may disappear. On the other hand, new sectors will emerge, including the sector for renewable energy and circular economy, creating new jobs.

The Netherlands government has developed and is still developing a wide range of policies, rules, and regulations to facilitate this transition, such as the National Climate Agreement, a major societal transition that aims to cut carbon emissions in the Netherlands in half by 2030. Another example is the policy vision on Circular Agriculture (LNV 2018), which presents a shared foundation for a societal transition towards circular agriculture and involves addressing agriculture, food (production and consumption), water, nature, climate, and the living environment in a series of concerted efforts. The question is, however, whether this vision will engender a true and radical transition or just several minor changes and efficiency improvements in the existing system. One of the more fundamental questions is whether a policy targeted at circular agriculture is compatible with the promotion of free trade in WTO and GATT negotiations and other fora, in particular by the European Union, the USA, and Japan, while at the same time practicing protectionism and subsidies for the domestic agricultural sector (Otero et al. 2013).

When it comes to radically new practices, insights, and values, however, small steps can resonate, ultimately bringing about large-scale changes (Bryson 1988). That is why, in response to the policy vision on Circular Agriculture (LNV 2018), Termeer (2019) advocates a 'small-wins' approach. This approach aims to work on major societal issues by dividing them into a series of 'small wins': small, meaningful steps with tangible results (Weick 1984; Vermaak 2009). The main thought behind this approach is that it keeps energy levels up and pushes forward progress in the transformation process without resorting to simplistic short-term gains or making impossible promises (Termeer and Dewulf 2017). By focusing on small-scale goals, people are less likely to be overwhelmed by the complexity of a given issue, which would restrict the freedom and precision of their thinking and increase the temptation for abstraction (ibid.). Responding to the 'small-wins' approach, however, Rotmans (2019) and Grin (2019) argue that this approach is too superficial and limited. According to Rotmans (2019), the dynamics of transitions include profound, broad, and slow changes as well as narrow and fast ones, with the essence of the transition being characterized by the 'parallelism of big and small, broad and narrow, fast and slow, construction and demolition'. Grin and Rotmans believe that particularly the transition towards circular agriculture requires system breakthroughs targeting a new culture, a new regime, a new paradigm, and a new infrastructure.

According to the ministry of Agriculture, Nature and Food Quality, the policy vision is not a blueprint for a system of circular agriculture, but rather describes a long-term process during which the government will have to adapt legislation, companies will have to apply circular principles and consumers will have to start paying fairer prices for their food (LNV 2018). Both producers and consumers will have to develop increased awareness and change their behaviour, forming two of the greatest challenges in realizing this agricultural transition. In response to the Ministry's future policy, Rotmans (2019), opposing Termeer (2019), argues that the government should not seek to direct matters: 'The harder the government pushes and pulls, the less room it leaves for other parties. Transitions originate from society and economy and though they can be facilitated by the government, this is all the government can do'. This viewpoint is not shared by everyone. A case in point is the historical Urgenda-court case on climate justice (see Sect. 2.3), where the Higher Court concluded, based on scientific evidence, that the Netherlands Government should step up its efforts to reduce greenhouse gases drastically by 2020.

Reasoning from the perspective of social contract theory (see Chap. 3), and in contrast to above argument of Rotmans (2019), government should do much more than facilitation only. After all, the purpose of the social contract is serving the common or greater good to ensure the sustainability of the society in question and protect the individuals within it. If not, political authority loses its legitimacy, and the social contract will be eroded, or will even fall apart, as illustrated by the Arab spring. Within the context of the sustainability transition, there is a wide variety of policy instruments and policy mixes that can be deployed for making a substantial contribution to the sustainability transition. For example, a systematic review of the European policy ecosystem shows that taxation is the most effective policy tool for mitigating unsustainable and unhealthy products and services in the food system (SAPEA 2020). Tax revenues, in return, can be used to provide positive incentives for realizing a transition to a sustainable and healthy agri-food system.

Many people feel that changing their individual lifestyle will not make a difference. You can put your best foot forward and install solar panels and insulate your house, or eating less meat or no meat at all, but the realization that a selection of only 100 companies is responsible for 71% of all carbon emissions since 1988 (CDP 2017) could be enough to discourage even the most optimistic mind.

The majority of Dutch citizens (65%) therefore believe that the government should take measures, according to a study by I&O Research (2019). When it comes to combating climate change, it appears that citizens are waiting for the government and businesses to lead by example (ibid.). Citizens believe having the government force businesses to adopt more sustainable production methods (62% have high expectations) and encourage technological development by these businesses (63%) will have the best effects. 6 out of 10 Dutch people (59%) agree with the statement 'As long as big companies fail to cut their carbon emissions, what I do will not make a difference'. The government has therefore come to realize that it must take action in all possible areas, including standardization, charges, subsidies, legislative amendments, binding agreements, budgetary choices, and green deals between public and private parties to facilitate the transformation towards sustainability. Also the introduction of a carbon tax for polluting businesses, as part of the National Climate Agreement, is a good example of government intervention.

Every change in society will provoke resistance, and attempts to change established patterns always come up against resistance, rigidity, and/or normative questions as to the legitimacy, justness, methods, and direction of the transition (Grin 2016; Meadowcroft 2009). Society is usually stuck in its old structures, thought patterns, and vested interests. Good intentions often fail because of the discrepancy between long- and short-term interests. Front-runners may introduce valuable initiatives, but they can struggle to establish a level playing field in the market. In addition, laws and legislation often offer insufficient scope for experimentation.

As a result, many efforts in the sustainability transition are struggling to move beyond the initiation phase and fail to realize acceleration and consolidation (see Sect. 4.1). This stalling of a transition can be characterized as follows (cf. Foresight and Commonland 2017):

- *'Fragmentation: many small, competing initiatives and isolated projects'*
- *'Narrow scope & lack of holistic approach: progress is measured on short timespans, and relative to competitors; transformation approaches focus on optimizing only a few dimensions, potentially at the cost of others; projects focus on only one link in the system'*
- *'Brittle: sustainability claims are based on marginal improvements'*
- *'Focus on inputs and processes rather than on outcomes'*
- *'Cause inflation: risk of losing credibility and being accused of greenwashing'.*

The sustainability transition will also give rise to unease, discomfort, and uncertainty, both financial and otherwise, which means resistance is inevitable (RNE 2016). To fight this resistance to change, mitigate negative consequences, and compensate for the adverse effects of the transition, it is very important to offer citizens and businesses appealing short-term or long-term prospects or attractors. The circular economy, for instance, will require more raw material collection, recycling and upgrading, creating new, low-skilled jobs in the process (ibid.). Large-scale sustainable development of the built environment will also create new jobs in the construction and installation industry (ibid.).

The sustainability transition is characterized by significant complexity and uncertainty. The sustainable development of our cities, for instance, has become such a complex and dynamic issue that it can no longer be tackled by just one party alone, such as government, private sector, or civil society (Karré 2018). The transition to climate-proof cities, for instance, will raise normative questions on what will make cities climate-proof and who should bear the costs involved in the process (Eriksen et al. 2015). Complex social issues of this kind are characterized by incomplete or fragmented knowledge and differing interests, values and ideas about problems, causes, and solutions.

The transition to a sustainable society will only succeed if everyone is given the chance to participate and if the costs, benefits, and risks are all shared fairly and proportionally. However, sustainable lifestyles are often restricted to the more affluent layers of society. Grants intended to encourage sustainable behaviour often flow to people with higher incomes, who can afford to invest in solar panels and an electric car. One of the major risks of the sustainability transition, therefore, is overemphasizing individual responsibility and relying too little on structural, systemic, and collective solutions. A major pitfall with problems such as climate change or sustainable consumption is that it tends to be reduced to personal choices and responsibility. In reality, however, these are structural problems requiring structural solutions, such as regulations, policy, and financial measures.

Besides an important role for government, it is clear that the sustainability transition can only succeed if all elements of society cooperate and assume their responsibilities. Sustainability certainly need not create a social divide between the 'green' and the 'grey' class, or between citizens and businesses, but should be part of an agenda for an inclusive society and social cohesion. A collective problem requires systemic, sustainable, and fair solutions, which requires all actors, including the government, the private sector, civil society, academia, and media to play their part.

2.5 What's Beyond the Sustainable Development Goals?

The term 'transition' or 'transformation' presupposes a fading to something new, a new state of mind, a new reality, a new normal, a new paradigm, or a new social contract. However, current literature on sustainability transitions is often occupied with the process, rather than its outputs, outcomes, or impacts (Köhler et al. 2019). This is illustrated by some transition scholars warning that describing the outcome can lead to processes in which the destination transcends the journey towards it (e.g. Haxeltine et al. 2016). Admittedly, precaution for trajectories where purpose justifies the means is undisputed, and requires careful attention for procedural justice (see Sect. 4.8). Moreover, dealing with complexity and uncertainty requires adaptive planning and governance (see Sect. 4.6), while a social learning process (see Sect. 4.7) is geared towards the process rather than a fixed goal (Bagheri and Hjorth 2007). This is something else, however, than developing a joint vision. A tangible and joint vision could serve as a vehicle to identify and develop shared and common values during the process of transformation. Agreement on these ethical and normative aspects is important for holding actor coalitions together during a transition process, and could be achieved through deliberation on shared beliefs and values, shared discourses, mutual understanding of common and diverging interests, procedural justice, and options for multiple value creation and mutual gains (see Sects. 3.8 and 4.8 in particular).

Within this context, literature on global environmental politics questions whether compliance with the Sustainable Development Goals should be considered as the ultimate goal of the sustainability transition, or whether we should be more explicit on what's beyond the 2030 Agenda (Wahl 2016; Dabelko and Conca 2019). Some scholars point out that the word sustainability itself is inadequate, as it does not tell us what we are actually trying to sustain (Wahl 2019). Wahl (2019) argues that design for sustainability is, ultimately, design for human and planetary health, which can be achieved through regenerative cultures (see Fig. 2.3).

Reicher and Hopkins (2001) argue 'that images of society's future are important for shaping social change. Social action must be animated by a vision of a future society, and by explicit judgements of value concerning the character of this future society' (cf. Chomsky 1970/1999, p100). A Natural Social Contract and the related concept of Transformative Social-Ecological Innovation, as proposed in this book (Chap. 3), serve as a vehicle to think about ways to improve current social contracts, targeting a sustainable, regenerative, healthy and just society, which can help

2.5 What's Beyond the Sustainable Development Goals?

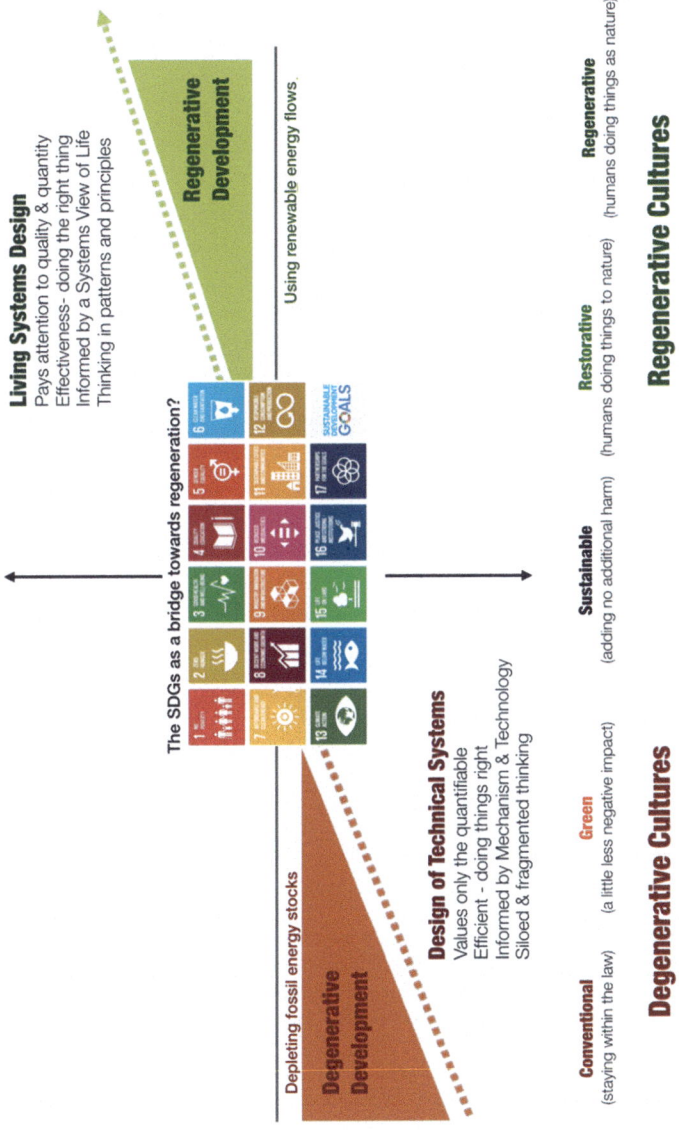

Fig. 2.3 Beyond sustainability: Designing regenerative cultures (based on Reed, 2006 & Roland, 2018)

policymakers, administrators and decision makers, concerned citizens, and professionals to make better decisions about how to organize our twenty-first-century society. In further developing ideas about the future of society, research on how people think about their own future may offer useful insights (Bain et al. 2013).

Open Access This chapter is licensed under the terms of the Creative Commons Attribution 4.0 International License (http://creativecommons.org/licenses/by/4.0/), which permits use, sharing, adaptation, distribution and reproduction in any medium or format, as long as you give appropriate credit to the original author(s) and the source, provide a link to the Creative Commons license and indicate if changes were made.

The images or other third party material in this chapter are included in the chapter's Creative Commons license, unless indicated otherwise in a credit line to the material. If material is not included in the chapter's Creative Commons license and your intended use is not permitted by statutory regulation or exceeds the permitted use, you will need to obtain permission directly from the copyright holder.

Towards a Natural Social Contract 3

In this chapter I will explain why and how the sustainability transition is humankind's search for a new social contract: a Natural Social Contract (conceptualization by author). I will start with a brief introduction on the origins of the social contract (Sect. 3.1), followed by a debate on the question whether there can be human progress without economic growth (Sect. 3.2) and a section on redesigning economics based on ecology, including circular and regenerative economies and cultures (Sect. 3.3). This chapter includes a debate on the role and scope of the free market (Sect. 3.4), as well as an examination of how the Anglo-Saxon and Rhineland models fare in this debate (Sect. 3.5). This chapter will also describe why we need a new social contract and what it should entail (Sect. 3.6). In doing so, I will embark on a quest for a Natural Social Contract (Sect. 3.7) and its theoretical foundations with multiple dimensions and crossovers (Sect. 3.8). This section concludes with an overview of fundamentals and design principles for a societal transformation towards a Natural Social Contract (see Table 3.4), which is a summary of Sect. 3.8 shaped as a course of action and is intended to help readers to grasp the core rationale of this book. For a better understanding of, and advancing the process towards, a Natural Social Contract this chapter presents a conceptual framework for Transformative Social-Ecological Innovation (Sect. 3.9), and how this will play out at various governance levels (Sect. 3.10).

3.1 What Is a Social Contract?

The sustainability transition constitutes a search for a new social contract. The core philosophy behind a social contract is that the members of a society enter into an implicit contract with the goal of living a better, safe life together (Kalshoven and Zonderland 2017). Such a contract includes agreements about public goods and services, for instance, as well as taxes, detailing how everyone contributes to and benefits from society. The contract describes the freedoms and obligations of all citizens: their rights and duties. This social contract does not exist in the sense that all citizens above the age of 18 sign a piece of paper. Rather, the social contract is an abstraction, a way of thinking that helps us understand how the world works that originates from the works of enlightenment philosophers (ibid. 2017).

Social contract theory has a long history in political philosophy. The main founders of classical contract theory are Thomas Hobbes (1588–1679), John Locke (1632–1704), Jean-Jacques Rousseau (1712–1778), and Immanuel Kant (1702–1804). Despite their differences, what these contract thinkers all have in common is that they tried to explain human society based on the idea that people once lived in some state of nature without rules and with unlimited freedom. In Hobbes's thinking, humankind naturally lives in a state of war (the conflict model), whereas Rousseau believed that humans were peaceful and timid in their pre-social state of nature, with social cohesion being created through consensus (the consensus model). According to Rousseau, the social contract enables humankind to pursue self-preservation by joining forces with others and sacrificing some individual freedoms for the will of the people. Rousseau used the metaphor of a contract to explore the relationship between individuals and their societies and legitimate government, and he argued that the ability to govern can only be legitimate if it comes from the people.

Following these enlightenment philosophers, contract thinking was given an important boost by the publication of A Theory of Justice, by social-liberal John Rawls (1971). There are also political philosophers, however, such as Michael Sandel and Charles Taylor, who criticize Rawls' work. Rawls does reserve a central position for the individual, for instance, but in Sandel's eyes fails to appreciate that all individuals are part of the community in some specific way (Sandel et al. 1985). Another important and more recent point of criticism is that 'Nature has had little or no intrinsic value for most (but not all) social contract theorists' (cf. O'Brien 2012), with no attention for the role of ecosystem services (Dobson and Eckersley 2006). The fact that ecological vulnerability translates into social and economic vulnerability, and a complex set of security and justice challenges, is an important omission in social contract theory.

In the past two decades, some scholars have argued that social contracts should be renegotiated due to the societal risks of climate change (O'Brien et al. 2009; Schellnhuber et al. 2011; Adger et al. 2013) and the ongoing ecological crisis (Jennings 2016), in particular given the co-evolving nature of risks and multi-actor influences on change (O'Brien 2012). Some scholars argue that the nature of environmental problems we face today requires new roles for states (Dryzek et al. 2002), while stressing several limitations of current social contracts: they can exclude those that may not recognize the legitimacy of government, and they can be influenced by non-democratic lobbying activities by powerful players (Weale 2011), and future generations are not represented. For instance, climate risks form a particular challenge for governments, given the related uncertainties and the often unequal distribution of risks and burdens (Pelling 2010).

A social contract is a more or less coherent whole of the freedoms, rights, rules and obligations that all residents have with regard to healthcare, education, labour, social security, and pensions, as well as in relation to our living environment, food, agriculture, nature, energy, water, the climate, and spatial planning. For example, all EU citizens have a right to the protection of fundamental rights, freedom of movement, and residence in the EU. The social contract might differ per country, but most European countries have similar rights and obligations as part of the social contract. Examples of such rules or obligations for EU citizens include compulsory insurance for medical expenses or compulsory education up to the age of 16. Likewise, all citizens are required to obtain a driving licence before driving a car and adhere to the traffic rules. Also pension schemes for employees, and phosphate rights for farmers are but some examples of the many arrangements in a social contract. The social contract, therefore, is key to the structure and functioning of our society. All citizens have a say in determining these arrangements through their voting rights, but there are more ways to give substance to a social contract. Each and every party in society can play a role in shaping and influencing the social contract, not only by means of our democracy (in various forms and at various levels), but also by bottom-up governance through civil society involvement, a participatory and inclusive society, transition management, multi-party collaboration, social entrepeneurship, corporate social responsibility, exercising the right to demonstrate, collective action, social innovation processes, citizen engagement, and through local, national, European or global citizenship (see Sect. 3.8 - social dimension of a Natural Social Contract). For each of these processes, it is necessary to identify how the governance of a societal transformation towards a Natural Social Contract can be designed, facilitated, and realized in effective and legitimate ways (see Sect. 3.8—institutional dimension). Attempts to change established patterns always come up against resistance, rigidity, and/or normative questions as to the legitimacy, justness, methods, and direction of the transition (Grin 2016, p. 112; Meadowcroft 2009). Hence, it requires inclusive procedures to broaden legitimacy of decisions and actions, through stakeholder participation and involving all layers of society. It also requires deliberation on shared beliefs and values, common interests, procedural justice, and opportunities for multiple value creation and mutual gains. In Sects. 3.8 and 3.9, as well as in Chap. 4, I will provide more detail on the governance approaches that are required for such a transition. In Sect. 3.6, I will continue the above discussion on why we need a new social contract and what it should entail (Sect. 3.6). Before doing so, let us start with a debate on the question of whether there can be human progress without economic growth (Sect. 3.2) and a section on redesigning economics based on ecology, including circular and regenerative economies and cultures (Sect. 3.3). This chapter also includes a debate on the role and scope of the free market (Sect. 3.4), as well as an examination of how the Anglo-Saxon and Rhineland models fare in this debate (Sect. 3.5).

3.2 Human Progress Without Economic Growth?

The social contract is not only about our rights and freedoms as stated in the constitution, but also about how we fairly distribute the costs and benefits of what we produce and consume in a country and about a broader definition of welfare (see

Sect. 2.1). It is clear, however, that a fair distribution of cost and benefits of what we produce and consume is not being achieved, since empirical studies show that inequality is increasing (Piketty 2013; Kremer and Maskin 2006). The assets are becoming more and more concentrated and a group of people is created that is extremely rich. On top of that, there is a well-established correlation between inequality and social and political instability (Russett 1964; Galbraith 2012; Stiglitz 2012, 2015). The problem, as Nobel Laureate Joseph Stiglitz argues, is that inequality can ruin democracy itself (Stiglitz 2012, 2015). Stiglitz argues that inequality is a choice—the cumulative result of unjust policies and misguided priorities (Stiglitz 2015).

Growing wealth inequality and the 2008 global credit crisis are merely symptoms of a deeper, systemic crisis. This can be traced back to decades of excessive production, consumption, and depletion of our natural resources and raw materials (Rotmans 2010). Over the past decade, a growing body of literature has been accumulating pointing out the contradiction between the pursuit of economic growth and ecological sustainability (Trainer 2011). We are now discovering that the ecological vulnerability translates into social and economic vulnerability, which is known as the paradox of prosperity (Sect. 2.1). For a more adequate conceptualization of the sustainability transition and the quest for a Natural Social Contract, we need a better understanding of the relationship between modern capitalist societies and the global ecological crisis. Naomi Klein, among others, has emphasized in 'Climate versus Capitalism' that the sustainability debate urgently needs to include a critical focus on economic systems (Klein 2015). Likewise, Mariana Mazzucato (2018) argues that we need to rethink capitalism, rethink the role of public policy and the importance of the public sector, and redefine how we measure value in our society, in particular since modern capitalist economies reward activities that extract value rather than create it (Mazzucato 2018). The literature on an alternative economy, written by economists such as Mariana Mazzucato, Paul Mason, Guy Standing, Colin Crouch, Eric Olin Wright, Paul Collier, and others, represents an expanding field of critical approaches to capitalism from various different angles. For instance, Paul Mason (2016) shows how the rise of the new digital economy is bringing about the decline of capitalism. According to Mason, capitalism cannot survive because primary resources (in particular information) are unrestrictedly available with an almost unlimited shelf life. This does not fit in an economic model based on private ownership. As a response, Mason argues for more cooperative schemes of free exchange—a 'sharing' economy to replace a predatory one— and more collective ownership as well. Likewise, Guy Standing (2019) argues for guarding our natural resources from private companies, by exploring the potential of the commons and commoning as an antidote against the erosion of society (see Sects. 3.8 and 4.3 for more information).

In particular since the 2008 global credit crisis the list with counter-proposals to unlimited economic growth has grown rapidly and is still counting. Many of these proposals are inspired by the 'Limits to Growth' report by the Club of Rome in 1972, followed by the UN Brundtland report on sustainable development 'Our Common Future' of 1987, and that led to the Millennium Development Goals dating from

2000, and eventually to the UN Sustainable Development Goals (2015), as part of the '2030 Agenda for Sustainable Development'. Although the 2030 Agenda has the ambition to end poverty and create a sustainable economic growth path and protect the planet from degradation, it does not state how to deal with the trade-offs and synergies of the various goals (Van Vuuren et al. 2017). For example, 'although some improvement with respect to global poverty can be observed, historical development patterns especially for environmental issues have mostly been at odds with this ambition' (cf. Van Vuuren et al. 2017).

On the most radical side opposed to unlimited economic growth there is a social movement and academic debate on degrowth, which started in the beginning of the twenty-first century. The English term 'degrowth' was 'officially' introduced at the 2008 conference in Paris on Economic Degrowth for Ecological Sustainability and Social Equity, which also marked the birth of degrowth as an international research area (Demaria et al. 2013). Kallis et al. (2018) review the broader literature relevant to degrowth debates.

The key propositions from this literature on degrowth are that economic growth is not sustainable and that human progress without economic growth is possible. More specifically, it argues that an equitable downscaling of production and consumption increases human well-being and enhances ecological conditions at the local and global level, in the short and long term (Schneider et al. 2010). According to Schneider et al. (2010) degrowth theorists and practitioners support an extension of human relations instead of market relations, demand a deepening of democracy, defend ecosystems, and propose a more equal distribution of wealth. Schneider et al. (2010) make an important distinction between depression, i.e. unplanned degrowth within a growth regime, and sustainable degrowth, a voluntary, smooth, and equitable transition to a regime of lower production and consumption (ibid.).

In addition to degrowth theorists there is burgeoning emerging literature, from diverse origins, with counter-proposals to unlimited economic growth. This varies from literature on **steady-state economy** (Daly 1973; O'Neill 2012; Kerschner 2010) to **green growth** (Ekins 2000; Hallegatte et al. 2011; Jänicke 2012; OECD 2011; European Commission 2011; UNEP 2011), to **circular economy** (Webster 2013, 2014; Ellen MacArthur Foundation 2015; EU Commission 2014; Murray et al. 2017; Prieto-Sandoval et al. 2018; and many others) and **regenerative economy** (Fullerton 2015; Moreno and Charnley 2016; Raworth 2017; Wahl 2016), with multiple definitions and distinctive developments in different contexts (Webster 2013; Lieder and Rashid 2016).

The green growth discourse has been the most dominant in the past 10 years, not in the least because the green growth concept was embraced by key global international organizations, including UNEP, the OECD, the European Commission, and the Global Green Growth Institute (OECD 2011; European Commission 2011; UNEP 2011) and eventually led to the adoption of the '2030 Agenda for Sustainable Development' by the UN member states in 2015.

More recently, however, the green growth discourse has been increasingly criticized, especially as economic growth is still a necessity in the proposed 'green' economies. First, a major criticism is that both neoliberal and keynesian economic approaches assume that prosperity stems from healthy GDP growth, but do not recognize biophysical limits to exponential growth, ignoring important lessons from ecology and thermodynamics about the natural limits of growth. In any case, this will reach a certain point where the incremental income is overtaken by the incremental damage, thereby decreasing global wealth (Hoffmann 2015; Fullerton 2015). To illustrate, a recent estimate shows that the 'hidden social and environmental costs' of the global food system and land use amount to USD 12 trillion, which is 20% more than the market value of USD 10 trillion (Pharo et al. 2019). Second, using GDP as the primary measure of our economic health does not accurately assess the economy or the state of the world and the people living in it (Van den Bergh 2017; Stiglitz 2019a, b). Third, there is a lot of criticism that the failure of market forces is solved by enlarging the market and introducing new market mechanisms (Fatheuer et al. 2015). This is done, among other things, by redefining the relationship between nature and the economy, in order to allow the market to arrange matters that were previously beyond its reach, such as pricing ecosystem services. This hides the many structural causes of the environmental and climate crisis. The result is a new version of the concept of nature as natural capital and the economic services of ecosystems, but it does not change the economic growth paradigm. New market mechanisms such as trading biodiversity credits or carbon credits in many cases do not prevent the destruction of nature, but only organize it along market lines (Fatheuer et al. 2015).

In a study by Van den Bergh (2017) 'agrowth' is proposed as an alternative to the disjunction between the 'green growth' and 'degrowth' positions. As it is impossible to know for sure whether growth and a stable climate are compatible, van den Bergh considers that it is better to be agnostic about growth (a-growth) and proposes a strategy that discounts GDP as an indicator, 'since growth is not an ultimate end, not even the means to an end' (Van den Bergh 2017). GDP is a measure of what the economy produces, but not for broader welfare. Nobel Prize laureate and pre-eminent economist Joseph Stiglitz points out that the interrelated crises of environmental degradation and human suffering of our current age demonstrate that 'something is fundamentally wrong with the way we assess economic performance and social progress'. He argues that using GDP as the chief measure of our economic health does not provide an accurate assessment of the economy or the state of the world and the people living in it (Stiglitz 2019a, b). By contrast, there are many non-monetary ways of measuring well-being (Mazziotta and Pareto 2013; Allin and Hand 2017; Fleurbaey and Ponthière 2019; Veneri and Murtin 2019; Hoekstra 2019). Many things of value in life cannot be fully captured by GDP, but they can be measured by metrics of health, education, political freedom, and metrics of sustainability, for example, to measure to extent of resource depletion (or circularity), pollution, energy use, climate change, biodiversity, ecosystem services, and so on.

Jeroen van den Bergh (2017) points out that 'green growth' is the dominant strategy among those accepting climate change as a serious threat and searching for solutions which minimize growth effects. Citing van den Bergh: 'The Paris climate agreement reflects this, through its voluntary national pledges without back-up from

3.2 Human Progress Without Economic Growth?

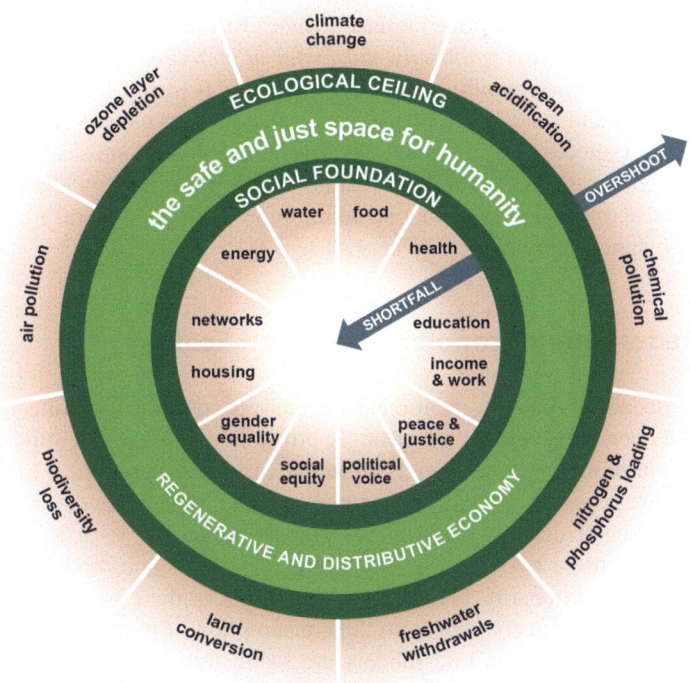

Fig. 3.1 Doughnut economy (Raworth 2017)

globally consistent policies. One must expect non-compliance, energy rebound and carbon leakage as a result, promising the agreement to be highly ineffective' (Quote from interview by AUB, 2017).

In Kate Raworth's book Doughnut Economics (2017) she argues that markets are inefficient and growth cannot continue unpunished. The carrying capacity of the earth must be respected and the economy must offer all people a decent life. Raworth uses the doughnut metaphor for this (see Fig. 3.1). The dough section of the doughnut represents a sustainable economy, the empty heart indicates the social deficits that may occur and marks the outer edge when ecological ceilings are exceeded. This means that the economy must adapt to the social and ecological preconditions, even if this would slow down economic growth. Between the social and planetary boundaries there is an environmentally safe and socially just, in short sustainable space within which humanity can flourish (Raworth 2017).

As such, her book is a counter-proposal to mainstream economic thinking that formulates conditions for a sustainable economy. Raworth calls for bringing 'humanity back at the heart of economic thought' (Raworth 2017). She argues that not everything can or should be left to the market, that the 'rational actor' model of economic conduct is problematic and that we cannot rely on the processes of growth to redress inequality and solve the problem of pollution. This plea echoes the work of many others, such as Nobel prize laureate Elinor Ostrom, who argued that *'neither the state nor the market is uniformly successful in enabling individuals to sustain long-term productive use of natural-resource systems'* (Ostrom 1990). According to Ostrom, the joint and sustainable management of commons cannot succeed without institutions for collective action (ibid.). Based on extensive empirical research she showed that common pool resources need not succumb to the so-called tragedy of the commons (exploitation by someone taking more than their share) if a system of checks and balances prevails (see Sect. 4.4).

3.3 Redesigning Economics Based on Ecology

In the past decade the above mentioned counter-proposals to unlimited economic growth have been subject to an increasingly broad debate in the scientific as well as the policy community. A wide variety of initiatives and programmes, from local to global level, are being elaborated, operationalized, and implemented, for example, in the form of circular economy and closely related concepts, such as regenerative economy. While these two concepts are not exactly the same, both with multiple definitions, the commonality between both concepts is their key proposition that wealth creation can be decoupled from the consumption of finite resources. Bottom line is that these new economic models are using the 'universal principles and patterns underlying stable, healthy, and sustainable living and nonliving systems throughout the real world as a model for economic-system design' (Fullerton 2015). Redesigning our industrial system of production and consumption around the circular patterns of resource and energy use that we observe in ecosystems is only one example of redesigning our economy using the insights of ecology (Wahl 2016).

The concept of the circular economy (CE) has become very popular in Europe and increasingly other global regions. It has been a catalyst at European policy level (Webster 2013) and has become influential across business circles (Howard et al. 2019). It has become so popular because it offers a solution that will allow countries, firms, and consumers to reduce harm to the environment and to close the loop of the product lifecycle (EU Commission 2014; Murray et al. 2017; Prieto-Sandoval et al. 2018), which stands in sharp contrast to the deeply rooted and intensive linear economic activity that is depleting the planet's resources (Prieto-Sandoval et al. 2018).

More than 100 different definitions of circular economy are used in scientific literature and academic journals. There are so many different definitions on Circular Economy because the concept is applied by a very diverse group of researchers and professionals (Kirchherr et al. 2017). Definitions often focus either on system change or on resource use. According to Korhonen et al. (2018), definitions that focus on system change often emphasize three elements, namely closed cycles, renewable energy, and systems thinking. Definitions that focus on raw material use often follow the 3R approach, namely 'Reduce' (minimum raw material use), 'Reuse' (maximum reuse of products and parts), and 'Recycle' (high-quality reuse of raw materials).

3.3 Redesigning Economics Based on Ecology

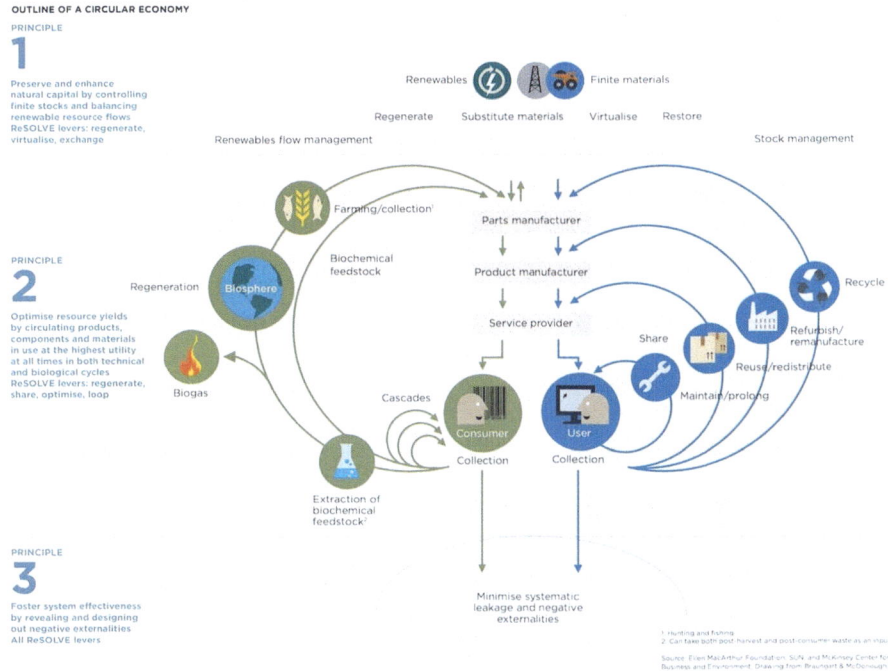

Fig. 3.2 Outline of a Circular Economy (Ellen McArthur Foundation 2015)

However, with a definition that only focuses on the use of raw materials, there is a risk that only optimization and efficiency improvements within existing systems are looked at, without the system itself being modified. Moreover, there is no guarantee that simple strategies of material recycling, as propagated by the various definitions of this concept, will lead to ecological sustainability (Desing et al. 2020) or to social justice and an inclusive society.

For example, some researchers argue that social inclusivity is a necessary part of the circular economy (Korhonen et al. 2018). In short, the transition from a linear economy to a circular economy does not only mean adjustments to reduce the negative effects of the linear economy. Rather, it is a systemic shift that builds long-term resilience, generates business and economic opportunities, and delivers environmental and social benefits (Ellen MacArthur Foundation 2015).

In the report 'Towards a Circular Economy', the Ellen MacArthur Foundation/ McKinsey schematically outlines the principles of the circular economy. See Fig. 3.2.

The British Standards Institution (BSI 2017) introduced six key principles of circular economy (for more details see Sect. 7.4):

1. System thinking: understand how your business impacts the whole ecosystem.
2. Innovation: manage resources for more value creation.

Fig. 3.3 Principles of a Regenerative Economy (Adapted from Capital Institute, 2020)

3. Stewardship: take responsibility for the ripple-effect impacts that come up from your business activities.
4. Collaboration: secure benefits at system wide level by strong cooperation in the value chain.
5. Value optimization: keep materials at the highest value and function quality.
6. Transparency: reveal to everyone the environmental impact of all your business activities.

In addition to Circular Economy, which is primarily focused on closing the loop of the product lifecycle, the model of Regenerative Economy is more holistic and explicitly builds on the natural design principles of healthy ecosystems (see Sect. 4.4). The principles of a regenerative economy are visualized in Fig. 3.3. Fullerton (2015) explains that '*a Regenerative Economy maintains reliable inputs and healthy outputs by not exhausting critical inputs or harming other parts of the broader societal and environmental systems upon which it depends'*. According to Fullerton, Regenerative Economy is '*a theory of political economy that transcends the contemporary debate between the neoliberal economics preferred by conservatives on the political right and the Keynesian economics generally preferred by liberals on the political left*' (ibid., page 40-41).

Taken together, the above debate marks an ongoing paradigm shift towards circular and regenerative economies and cultures. In Part III of this book I will apply these new economic models, among others, within the context of the transition to a sustainable and healthy agri-food system (Chap. 6) and urban sustainability transitions (Chap. 7).

In the following sections I would like to draw attention to the consequences of this ongoing paradigm shift for existing social contracts. It requires a debate on the role and scope of the free market and a critical reflection on the Anglo-Saxon and Rhineland model which currently determine the organization and functioning of capitalist societies.

3.4 Debate on Role and Scope of the Free Market

Since the 1970s, the market economy in many countries, mostly Western countries, has silently developed into a market society (Sandel 2012), which means the way people live together is dominated by individualism, self-interest, the free market, and a focus on profit and economic growth, shifting social and ecological values and interests to the background. Since the 1970s, and the following decades, many Western countries were too easily involved in the story that if the market arranges it, it is by definition better and more efficient.

Prof. Dr. Kim Putters, Director of the Netherlands Institute for Social Research (SCP), explains the problem of a market society as follows (AWVN 2016): 'The privatisation policy pursued by the government in recent decades has made a significant contribution to over-individualisation. Citizens have been turned into customers, but customers behave differently from citizens. Customers demand what they have paid for but feel no duty to the community. If people do not feel they are co-owners of the community's collective goods, such as the local park or their neighborhood, there are no collective standards. As a result, they never question what effect their behaviour has on the community'. Instead, citizens have increasingly started deriving their identity, security, and social status from consumption. A general passion for excess and materialism has developed, as well as a focus on beautiful, luxurious, and expensive items.

The triumphal march of the free market began in the early 1980s, after which the shift from a market economy to a market society has seen Western society become gradually more focused on individualism, the free market, and economic growth. Academic literature (see Rojas 2014 for an overview) emphasizes that our capitalist market society is an ideal foundation for consumer society, resulting in excessive mass consumption, environmental pollution, and depletion of natural resources. A capitalist market society usually rewards activities that extract value rather than create it (Mazzucato 2018). According to economist Mariana Mazzucato (2018) we will need to rethink capitalism, rethink the role of public policy and the importance of the public sector, and redefine how we measure value in our society (Mazzucato 2018).

A market economy is a valuable, effective tool for organizing productive activity. However, when market values permeate every aspect of human lives, and when market economy silently develops into a market society, in which social relationships are formed in the image of the market, it may become problematic (Sandel 2012). In his book 'Not everything is for sale' (2012), American political philosopher Michael Sandel points out the disadvantages of a society run by market values. For the majority of policies, such as those related to climate change, agriculture, water management, nature, biodiversity, education, healthcare, housing, transportation, it is important to engage in a debate on the role and scope of the free market. Sandel identifies two important drawbacks of a market-based society: (1) inequality and (2) corruption. Sandel explains that in a society where everything is for sale, people with few resources struggle more than people with ample resources, which leads to inequality. The more that money can buy, the more important money or the lack thereof will become. This is particularly problematic when money also buys access to social services, such as education and healthcare. This is also a problem within the sustainability transition, especially if sustainability is mainly framed as a matter of ethical consumption. It is easier for people with ample financial resources to spend money on sustainability than people who are not as wealthy. In that case, the moral choice will always be the more expensive choice. The risk is that this may create separate worlds: a 'green class' versus a 'grey class', where ecological and economic inequality go hand in hand. As a result, the gap between the rich and poor is widening, further exacerbated by the commercialization of social life (Sandel 2012). The second reason Sandel (2012) identifies is the corrupting effect of the market. Putting a price tag on everything opens the door for corruption, as it facilitates abuse of power and money. This is why we need a public debate as to where market forces should and should not play a role. Society will have to determine how important aspects of life such as healthcare, education, family life, nature, food, art, and civic duties should be valued.

Citizens cannot rely on the market to solve social problems such as social inequality, climate change, and other environmental problems. In his book Economics for the common good, Nobel laureate Jean Tirole explains that the best solution for climate change is putting a price on greenhouse gases. This, Tirole believes, is the only way, although he does add a careful analysis of why this solution will not be successful either (Tirole 2017).

Take, for instance, the 1997 Kyoto Protocol, which aimed to reduce global greenhouse gas emissions by introducing various new market mechanisms, backed by industrialized countries:

1. Emissions trading: a country or polluter with more carbon emissions can buy the right to emit more from a country or party that emits less.
2. Clean development mechanism (CDM): industrial countries can finance emission reduction projects in developing countries, such as by providing financial

3.4 Debate on Role and Scope of the Free Market

support to forestry or soil protection initiatives. In exchange, they receive credits (carbon credits) that allow them to emit more themselves, so that they do not have to cut emissions in their own country.

In a nutshell, countries or polluters can use financial injections in developing countries to buy permission to pollute. According to critics, these measures make the Kyoto Protocol a failed agreement (Gilbertson and Reyes 2009; Rosen 2015). Even the World Bank, a proponent of carbon trading, highlights the various shortcomings of the CDM (World Bank 2010).

There is a risk of introducing similar market forces when pricing the so-called ecosystem services. An ecosystem service is a benefit provided to people by an ecosystem, including production services (food, water, wood), regulating services (air purification by greenery and forests, water treatment by wetlands, pest control, pollination, water and climate regulation), cultural services (recreation, healthcare, cultural history, inspiration and religion), and support services (soil formation, primary production, the nutrient cycle, and biodiversity). The advantage of putting a price on these ecosystem services is that it generates awareness, especially since these services are often invisible, public goods. However, there is a risk that pricing ecosystem services can bring market forces into play, turning natural capital into an asset that can be appropriated, used, and traded by humankind. To present this from happening, regulation is required that curbs the adverse effects of market forces and guard ecological boundaries. To open the door for regulation, though, necessitates a public debate about the role, scope, functioning, and consequences of the free market and the associated norms and values. For example, 'since climate change and other threats occur over multiple scales and across the very long run, they demand governance and tools like incentives and feedback loops that act as guard rails and, where necessary, limits that coordinate across scales and focus on the long-term' (cf. Fullerton 2015).

In the past two decades, there has then been a move in environmental economics to regard such things as natural capital and ecosystems functions as goods and services. (e.g. Millennium Ecosystem Assessment 2005). However, this is far from uncontroversial within ecology or ecological economics due to the potential for narrowing down values to those found in mainstream economics and the danger of merely regarding Nature as a commodity. This has been referred to as ecologists 'selling out on Nature' (McCauley 2006).

In short, more oversight and adequate regulations alone are far from sufficient to curb the adverse effects of market forces. The financial crisis, banking crisis, and debt crisis are merely symptoms of a deeper, systemic crisis. This can be traced back to decades of excessive production, consumption, and depletion of our natural resources and raw materials (Rotmans 2010). Today's economy was not designed to cope with the perverse effects of such practices and it therefore lacks long-term feasibility. Without a transition to a sustainable economy, we will automatically create the next crisis (Grin et al. 2010).

3.5 Anglo-Saxon Model Versus Rhineland Model

Political analysts believe that the 2008 global credit crisis ushered in the bankruptcy of financial capitalism. When it became clear that banks had been taking uncontrolled, irresponsible risks that put the global economy in serious jeopardy, many European countries expressed their support for more government control over the financial sector.

Ever since, the Rhineland model seems to have been gaining popularity, with the focus shifting to include other stakeholders (the Rhineland model) than shareholders alone (the Anglo-Saxon model). To put it simply, the Rhineland model is the economic system traditionally used in the countries that the Rhine flows through (Switzerland, Germany, France, and the Netherlands) and related economies (the Scandinavian countries, Belgium, Luxembourg, and Japan). This model rests on a consultative culture, solidarity, an appreciation of craftsmanship and values other than money, such as quality and happiness.

The Rhineland model is diametrically opposed to the Anglo-Saxon model in its raw form, as can be found in the USA, Great Britain, and Singapore, where money often seems to be the only measure, shareholders have the final say, and processes must be managed as efficiently and cheaply as possible (Schouten and Spijker 2017).

These two models were identified and defined by Frenchman Michel Albert in his classic book Capitalism vs. Capitalism, published in the early 1990s. Bakker et al. (2005) highlight the key differences between these two models in Table 3.1.

The Rhineland model is typically associated with the social market economy that developed on the European content after the Second World War, especially in Germany, the Benelux, and Scandinavia. Ludwig Erhard, Germany's first Federal Minister of Economic Affairs (1949–1963) is considered the founder of the social market economy, which aims to achieve the greatest possible prosperity while providing the best possible social protection. The social market economy is seen as a process and had to continuously adapt to new circumstances, such as globalization, digitization, climate change, ageing populations, and migrations.

Since the 1990s, the Rhineland model has come under increasing pressure from Anglo-Saxon values, such as the highly instrumental approach to the free market, shareholder value, human capital, management/control, and accountability (Goodijk 2009). In the Netherlands, market thinking also crept into sectors that were traditionally regulated by the government, such as utilities, the post, public transport, and even healthcare.

This shift to a more Anglo-Saxon value system was in line with a major economic-political turnaround set in motion in the 1970s, which saw the market gain ground in a great many policy areas as state power diminished (Zuidhof 2014). This development is called neoliberalism by some, but its supporters in the Netherlands prefer terming it classical liberalism, inspired by the work of Friedrich Hayek and Milton Friedman. To illustrate Hayek's influence, Dutch Prime Minister and Liberal Mark Rutte has written several essays about Hayek, including in Elsevier (2007) and Vrij Nederland (2011), and has repeatedly highlighted the influence of Hayek's ideas on his politics. In general, since the 1970s the dominant form of

3.5 Anglo-Saxon Model Versus Rhineland Model

Table 3.1 Differences between the Anglo-Saxon and Rhineland models (Translated and adapted version, based on Bakker et al. 2005)

Model Aspect	Anglo-Saxon	Rhineland
Focuses on	Short-term profits: • Shareholder value. • Money is power. • May the best man win. • Win–lose. • You are either with us or against us.	Continuity and trust: • Customer satisfaction. • Employee satisfaction. • Shareholder satisfaction. • Win–win. • A more multi-faceted approach to relationships.
Dominant philosophy	Financial thinking	Industrial thinking
Performance focused on	Next quarter	Continuity
Acquisition	Power to capital	Protective structures
Organization is	Supportive 'money-making machine'	A working community, a 'necessary evil' for realizing complex products
Central focus	Money, power, and heroism	Expertise, substance
Manager is	An MBA, because managing is a skill	A hands-on foreman (cf. Guilds and their Master's test)
Expertise is	The responsibility of employees	The responsibility of employees and the organization
Focus on	Personal utility	Personal dignity
Motivation	Extrinsic (money, incentives)	Intrinsic (nature of the work)
View of mankind	Mechanistic	Humanistic
Remuneration depends on	Productivity	Position
Labour	Is a cost factor	Has a social component
Employees	Input	Embody the organization
Funding through	The stock market	Banks and family
Companies	Mostly public	Various business models
National	Minimum government interference: the market governs (the invisible hand)	Government plays an active role. Consensus between employers, employees, and financiers
Relies on	Military/hard power	Economic/soft power
Unilateral/multilateral	Unilateral	Multilateral
Central focus	Individuals	Mutual relationships
Leading principle	Individual success: the American Dream	Collective well-being, culture, open, and 'feminine'
Attitude to minorities	The winner takes it all	Minorities get a share
Bankruptcy is	The start of something new	The end
Focused towards	The USA	Asia

(continued)

Table 3.1 (continued)

Model		
Aspect	Anglo-Saxon	Rhineland
Driven by	Technology (D&E) and market	Design and science (R&D(&E)
The Netherlands	Country of traders: Merchants Preachers	Country of innovation: Ahead of the socio-economic curve Ahead of the technological curve
Education funded by	Government funds public schools	Government also funds private schools
Coordinated by	Rules	Shared values
Character	Adventurous, exciting, and passionate	Cautious, careful, prudent, boring
Relationship between businesses	Competition	Cooperation

capitalism is generally called neoliberalism and the Washington Consensus, with heavy influence from the free-market-oriented Chicago School of economics and the Hayek philosophy (Fullerton 2015). Since the 2008 global credit crisis, a resurgence of Keynesian and particularly post-Keynesian ideas has pushed back into the mainstream debate, calling for a greater role for the State in regulating free market capitalism (Fullerton 2015; Biebricher 2017).

Neoliberalism, or classical liberalism, is difficult to define, firstly because it consists of three major movements which each have a different view of the relationship between the economy and democracy and between the market and the state (Biebricher 2017). These movements disagree on the desirability of government intervention, though they do all agree that the government is responsible for creating the underlying conditions that the free market requires in order to flourish. The belief in the infallibility of the market as the ultimate truth teller is at the heart of neoliberalism, whereas opinions differ about whether this requires big or small government and what exactly the domain of government intervention is.

Ronald Reagan and Margaret Thatcher, the political figureheads of neoliberalism in the 1980s, started the process of liberalization with tax cuts and reining in the welfare state. After the fall of the Berlin wall in 1989, the capitalist camp was able to proclaim its political and socio-economic superiority. According to Francis Fukuyama, the global battle between capitalism and communism was won (Fukuyama 1992). Capitalism was declared the victor after the fall of communism, but what did that really mean? Which type of capitalism had won? Was it Reagan's and Thatcher's liberal capitalism or the type of capitalism practised in Germany and the Netherlands, among other countries? Had the Anglo-Saxon model or the Rhineland model triumphed (Bakker et al. 2005)? Sandel (2012) believes that, over the past decades, our market economy has, consciously or subconsciously, made way for a market society, dominated in many respects by the Anglo-Saxon model.

In response to the consequences of the Anglo-Saxon model, such as shareholder capitalism, self-regulation, and excessive individualism, supporters of the sustainability transition are now looking for answers in new incarnations of the Rhineland model (see, e.g. Rotmans 2010; Latour 2018; Varoufakis 2019). The leading principle here is striving for a balance between government, the market, and civil society. Rotmans (2010) describes the renewed Rhineland model as follows: 'Anglo-Saxon values will make way for more harmonious relationships between the market and government, between labour and capital and between individualism and capitalism. The renewed Rhineland model is based on fundamental moral values such as well-being, quality, ecological preservation, long-term thinking, and "togetherness". These are all prerequisites for a sustainable economy. From a social point of view, this means, among other things, that professionals such as police officers, nurses, and teachers should not be regarded as mere production factors. They are professionals who must be able to develop and be rewarded for their performance. The slow decision-making process of the traditional Rhineland model, consulting with myriad interest groups, will be swapped out for effective cooperation with new coalitions in horizontal networks'.

3.6 Looking for a New Social Contract

In 2020 the Coronavirus pandemic shakes the world to its foundations and will probably create a new reality once the pandemic has been tackled. The weaknesses of a market-based society, primarily focused on economic growth and ever-increasing circulation of goods and people, have been painfully exposed. First, the people working in health care, food production, education, and the police, who were regarded as mere production factors in a neoliberal model and who had to hold up their hands for a decent salary before the Corona-crisis, are now the heroes of society. Of course they already were, but appreciation for these professionals failed to materialize because it was overshadowed by the over-appreciation of the free market, market-based values, privatization, and unlimited economic growth. Second, another weakness exposed by the Corona pandemic is that a model of unlimited circulation of goods and people has substantially fueled the spread of disease around the world. Third, climate change promotes the emergence of serious disease outbreaks (Laaksonen et al. 2010; Rees et al. 2019), and the coronavirus will certainly not be the last causing large-scale societal disruption.

Before the Corona-crisis, the world witnessed a surge of massive civil protests in many countries in 2019, including Hong Kong, Chile, Bolivia, Ecuador, Guinea, Haiti, Honduras, Nicaragua, Malawi, Russia, Sudan, Zimbabwe, Egypt, Algeria, Iraq, and Lebanon, as well as in Europe, for example, in the UK, Spain, France, Germany, and The Netherlands. The trigger of these protests varied from country to country, but there are a number of underlying commonalities, all of which can be described as threats to the social contract. The common factors pushing people to protest include corruption, inequality, cost of living, climate justice, and political freedom, while in 2020 protests over racial injustice and police violence dominated the news.

One of the standout moments came in September 2019, when more than 7.6 million people took part in a week of climate strikes in 185 countries. The climate protesters demanded urgent action on the escalating ecological emergency, and some argued that politicians and governments had 'broken the social contract' by not anticipating and responding adequately to the climate crisis. Indeed, the purpose of the social contract is serving the common or greater good to ensure the sustainability of the society in question and protect the individuals within it. In other words, the social contract is expected to provide security and justice for all. If not, political authority loses its legitimacy, and the social contract will be eroded or will even fall apart, as illustrated by the Arab spring. Climate change and its effects are inextricably linked to complex questions of security and justice (see Huntjens et al. 2018) and therefore relates directly to the social contract.

The rising threats to the social contract on a global scale have led UN Secretary General Guterres to sound the alarm on 25 October 2019, urging leaders everywhere *'to listen to the real problems of real people'*. He also stressed that the world *'needs action and ambition to build a fair globalization, strengthen social cohesion, and tackle the climate crisis'*.

The question is, therefore, whether current social contracts can still provide an adequate response to the challenges of the twenty-first century, such as corruption, inequality, climate change, pollution, and depletion of natural resources. In recent years there have been several proposals to account for climate change (mitigation and/or adaptation) in current social contracts (e.g. see O'Brien et al. 2009; Schellnhuber et al. 2011; Adger et al. 2013, Jennings 2016). One such proposal is developed by the German Advisory Council on Global Change (WBGU) in 2011 (Schellnhuber et al. 2011), addressing the unsustainability of our current carbon-based economic model, and as response, the WBGU-report proposes a new social contract based on a transformation towards a low-carbon society. In the field of climate change adaptation, O'Brien et al. (2009) and Adger et al. (2013) propose to renegotiate social contracts between citizens and states as a primary mechanism for adaptation. Jennings 2016) provides an ethicist's reckoning with how our political culture, broadly construed, must change in response to climate change.

In this book I fully endorse above proposals, but also challenge them, given their limited focus on either climate change mitigation (e.g. Schellnhuber et al. 2011) or climate change adaptation (e.g. O'Brien et al. 2009; Adger et al. 2013) or a too structuralist point of view (e.g. Jennings 2016), with limited attention for how to realise change on the ground. In taking an actor-centric institutional perspective (see structure-agency debate in Sect. 3.9), my book addresses two mistakes: a too structuralist point of view (common in political economy) and voluntarism (common in actor-centric research of specific innovations). A broader and more existential societal transformation is required for asserting a sustainable and healthy future. It is true that most people are gradually realizing that global warming is a major problem, but is climate change not just a symptom of a deeper, systemic crisis? A new social contract should be able to respond to societal fault lines, the most comprehensive of which is the divide between humans and nature, as well as the deeply embedded capitalist economic logic, resulting in increasing wealth inequality as well as market-based societies where citizens have been turned into customers and consumers and demand what they have paid for, but feel no duty to the community or environment. In current anthropocentric social contracts, natural resources are viewed to be used exclusively by humans, to serve the needs of humanity and the needs of our current economic systems with a singular focus on economic growth. The loss of biodiversity, environmental degradation, bio-industry, land, water, and air pollution, and fossil energy consumption, for instance, show that the way we deal with nature is profoundly disturbed. However, most of the scientific frameworks for sustainability transitions and transformation research remain limited by not reflecting on how deeply embedded the divide between humans and nature in combination with the capitalist economic logic has become in our current social contracts. Therefore I argue for a fundamental shift from our current anthropocentric and economic growth-oriented social contract towards a more ecocentric, regenerative, and natural social contract (see Sect. 3.7), which does justice to the natural position of humankind and society within a larger ecosystem, that of planet Earth.

Thinking about ways to improve the social contract, targeting a sustainable, regenerative, healthy and just society, can help policymakers, administrators and decision-makers, citizens, and professionals to make better decisions about how to organize our twenty-first century society.

3.7 A Natural Social Contract

The sustainability transition implies a large-scale societal transformation towards a Natural Social Contract, in which Transformative Social-Ecological Innovation (TSEI) will be needed in different fields and at different levels of scale (see Sect. 3.10 for examples).

The five decades from 2000 to 2050 will go down in history as the sustainability transition, by some called the 'Great Mindshift' (Göpel 2016) or the next 'Great Transformation' (Schellnhuber et al. 2011), referring to a redirection of civilization that recalls the advent of market economies described by Karl Polanyi in *The Great Transformation* (Polanyi 1944; Haberl et al. 2011; Leggewie and Messner 2012). The transformation to a sustainable, healthy, and just society is humankind's quest for a new social contract—in this book coined as a Natural Social Contract—which requires a fundamental shift from our current anthropocentric social contract towards a more ecocentric and regenerative social contract, acknowledging society as a social-ecological system (see Sect. 3.8). A Natural Social Contract is an unavoidable and logical response to the most comprehensive societal fault lines of our times, which can be traced back to two common denominators. First, the schism between humans and nature and the dominant anthropocentric world view that arose during the Enlightenment. Second, the capitalist economic logic, in particular the unsustainability of infinite economic growth in a finite world, and the belief in the infallibility of the free market, that arose after the Second World War. In particular since the 1970s, and the following decades, many Western countries were too easily involved in the story that if the market arranges it, it is by definition better and more efficient (Sandel 2012).

A societal transformation towards a Natural Social Contract constitutes an existential change in the way humankind lives in and interacts with its environment, in harmony with nature, and focusing on people as members of a community and as part of a natural ecosystem. It does justice to the natural position of humankind and society within a larger ecosystem, that of planet Earth. A Natural Social Contract enables humankind to pursue self-preservation and higher levels of well-being by joining forces with others and with nature while at the same time putting an end to unlimited economic growth, overconsumption, and over-individualization, for the benefit of ourselves, our planet, and future generations. Key differences of a Natural Social Contract compared to existing social contracts are described in Table 3.2.

Seeing humankind as part of a natural ecosystem is the opposite of the Western dominant social paradigm of a firm boundary between humanity and the environment. Bruno Latour (2012) argues that 'the essence of the "modern constitution" lies in the fiction of an ontological separation between humans and society on one side and nature and non-humans on the other side'. In particular during the Enlightenment—from the seventeenth to nineteenth centuries—the separation between nature and human society became more dominant due to rationalists such as Descartes and Bacon, as well as Enlightenment thinkers such as Newton, Kant, Adam Smith, and Montesquieu. For example, René Descartes' (1596–1650) made a strict separation between humankind and the rest of the universe by claiming that only humans have a 'spirit'. The rest of nature, he believes, is only matter. Descartes' radical division is particularly important because it is such a clear representation of the dominant anthropocentric world view that arose during the Enlightenment (Mommers 2019). Descartes' main contribution is that he applies a varnish of rationality to the hierarchy of people in nature. In post-Enlightenment Europe, the idea that humankind is superior to nature is no longer an opinion; it is upgraded to a fact. If you disagree, you are not considered rational. This, effectively, set the schism between 'society' and 'nature' in stone (Patel and Moore 2017). As a result, 'modern societies have engaged in increasingly disruptive modes of interaction with the biophysical environment, and this is widely perceived as not simply a side effect, but a characterizing trait of modern societies' (cf. Jackson 2009). During this 'modernization' process in the past centuries, where humanity aspires to transcend nature and to control the external world, modern societies have also lost sight of some basic principles for life on earth in the long term and thus ignoring vital benchmarks for a sustainable and healthy society (see Sect. 4.5: design principles from nature). Contemporary scholars emphasize that natural conditions are not separate from social processes (Berkes et al. 2000; Anderies et al. 2004; Skandrani 2016). Within a Natural Social Contract, the term 'social-ecological system' is used 'to highlight the integrated concept of humans within nature and to address the delineation between social and ecological systems as artificial and arbitrary' (cf. Berkes et al. 2000).

3.7 A Natural Social Contract

Table 3.2 Natural Social Contract compared to existing social contracts

	Liberal/Interest-based	Social/Right-based (negative)	Social/Right-based (positive)	Natural Social Contract
Key proponents	Hobbes, Hayek, Nozick, Gauthier, Buchanan	Locke	Rouseau, Kant, Rawls, Scanlon	Conceptualization by author
Development paradigm	Neoliberalism; government is responsible for creating the underlying conditions that the free market requires in order to flourish	Inclusive liberalism	Social democracy	Various forms of democracy, broad welfare, circular, and regenerative economies and cultures
Core policy concerns	Security and economic opportunity	Security and economic opportunity	Empowerment and equity	Human security, social justice, and planetary health
World view	Human-centred/anthropocentric	Human-centred/anthropocentric	Human-centred/anthropocentric	Social-ecological systems/ ecocentric
Core values	Free market trade, deregulation of financial markets and corporate sector, privatization of the public sphere, individualization, and the shift away from state welfare provision	Locke's fundamental natural rights: life, liberty, and property. These rights are referred to as 'negative freedoms', because they are the freedoms not to be restricted by another person	Freedom, equality, justice, and solidarity. It is not only about securing negative freedoms, but freedom has financial and social preconditions. Freedom must be legally ensured and effectively guaranteed (e.g. through education)	Respect and care to all life (including ecosystems as a whole), solidarity and togetherness (as being central to group life), environmental protection, and the sustainable and joint management of shared resources, collective well-being, democracy, and justice
Overarching goal of the social contract	Protection; maintenance of order	Protecting existing property rights	Promoting justice	Broad welfare through human security, social justice, and planetary health
Vision of individual	Rational actor; motivated by subjective ends. Individual as isolated from others	Rights-bearing citizen	Impartial actor, motivated by impersonal aims. Individual in relation to others	Human life is group life. Individual in reciprocity with social and natural environment and related stewardship

(continued)

Table 3.2 (continued)

	Liberal/Interest-based	Social/Right-based (negative)	Social/Right-based (positive)	Natural Social Contract
Vision of society	Individualistic, merit-based notion of justice	Individualistic	Commonwealth, equality-based social justice	Society as a social-ecological system. Humankind as a component of a greater ecosystem that of planet earth
Basis for social relations	Utilitarian	Mutual respect	Mutual respect	Mutual respect, solidarity, connectivity, social and environmental stewardship
View on the natural environment	Ecosystem is a black box. Natural resources to be used exclusively by humans, to serve the needs of humanity	Ecosystem is a black box. Natural resources to be used exclusively by humans, to serve the needs of humanity	Ecosystem is a black box. Natural resources to be used exclusively by humans, to serve the needs of humanity	Earth is the whole of which humans are subservient (but impactful) parts. Institutional and economic design based on lessons taught by nature

The overview of existing social contracts is a modified version (by author) based on Hickey (2011) and expanded with the Natural Social Contract (conceptualization by author).

3.8 Dimensions and Crossovers Within a Natural Social Contract

Every society, and thus every social contract, consists of several dimensions, including an economic, social, ecological, and institutional dimension. Each of these dimensions consists of a multitude of interconnected heterogeneous components. These dimensions themselves are a complex, dynamic, self-organizing, and evolving entity, so the four dimensions together lead to an enormous complexity (Spangenberg 2005). Change or problems in one dimension thus affects all dimensions and vice versa. Finding leverage points alone is not enough; systemic change also requires good insight into the interrelationships, for example, via (non-linear) feedback loops, and how the desired outcome can be achieved with maximum synergy effects and minimal 'trade-offs' (Kennedy et al. 2018). The connections between the dimensions must enable permanent coevolution, when working on transformative change, but also means a high degree of path dependence, with choices from the past determining the current structure. This path dependency is a reason for institutional stability, since institutional pressures force organizations to adopt similar practices or structures to gain legitimacy and support (DiMaggio and Powell 1983, 2000), and these institutions become firmly rooted in taken-for-granted rules, norms, and routines (Seo and Creed 2002). A societal transformation will always be a battle to overcome vested interests, change existing systems and paradigms. This explains, among others, why major societal transformations take on average about 30 years, which is also a realistic timespan for some of the fundamental systemic changes required for a Natural Social Contract.

A Natural Social Contract as proposed in this book is an open and broad theoretical framework across multiple dimensions (i.e. social, ecological, economic, and institutional) that serves to start a dialogue about ways to improve the current social contract, targeting a more sustainable, regenerative, healthy, and just society. It can help policymakers, administrators and decision-makers, concerned citizens and professionals to make better decisions about how to organize our twenty-first century society. It is an open framework in the sense that it is open for additional predictors and moderators in every dimension if they have a documented effect. This theoretical framework is visualized in Fig. 3.4. A clear delineation between dimensions is not possible, given the broad scope and overlap of some of the theories mentioned, while the positioning of a theory or concept in one dimension is sometimes artificial (for the purpose of visualization), given its relevance for other dimensions. The theories and concepts included are described in various chapters of this book and brought together in this theoretical framework, although 'theoretical context' is perhaps a better description. In any case, the purpose is to give a semi-structured overview of relevant theories and concepts that provide a better understanding of a Natural Social Contract. Each dimension, and possible crossovers with other dimensions, will be briefly described below (with reference to sections in this book for more detail).

Fig. 3.4 Theoretical and open framework for a Natural Social Contract. The inner circle represents the theoretical core, and the outer circle represents the theoretical context. A clear delineation between dimensions is not possible, given the broad scope and overlap of some of the theories mentioned, while the positioning of a theory or concept in one dimension is sometimes artificial (for the purpose of visualization), given its relevance for other dimensions

This section concludes with an overview of design principles for a Natural Social Contract (see Table 3.3), which is a summary of Sect. 3.8 shaped as a course of action and is intended to help readers to grasp the core rationale of this book.

Social Dimension

Relevant for a Natural Social Contract is an explicit emphasis on a human being's natural state as a social animal living in families, as a member of a group, community, collective, organization, or company. Human life is group life (Chambers

2018), regardless of whether these groups are large or small. In order to pursue self-preservation and higher levels of well-being, people depend on group life, and thus reciprocity with their social and natural environment and related stewardship. Within this context, 'sharing' is an important evolutionary trait of humans. 'Shared efforts allowed our ancestors to band together to hunt, farm, and create shelter, and reciprocal forms of altruism arose naturally from repeated interactions in such collective groups' (cf. Agyeman and McLaren 2017). This sociological aspect of a Natural Social Contract is diametrically opposed to the individualistic nature of the 'Homo Economicus' (more information below) in a market-based neoliberal society in which citizens have been turned into consumers and customers. But customers behave differently from citizens, since customers demand what they have paid for but feel no duty to the community or natural environment. 'The human evolutionary trait of sharing has largely faded in capitalist societies, due to commercialization of the public realm, rapid economic and technological change, and the rise of competitive individualism' (ibid.). In a Natural Social Contract, however, all individuals are considered to be part of the community in some specific way, in varying degrees and often in diverse roles. If you ask people what guiding principles they find important in life, helping others and protecting the environment come out strongest (at least in an analysis of ca 44,000 responses from people representing 22 European countries and Israel (Bouman and Steg 2019). In Table 3.3 I provide an overview of several important aspects of human nature that could be re-invigorated through a Natural Social Contract.

With regard to people's reciprocity with their social and natural environment I make a distinction between 'social' and 'natural' environment. People's reciprocity with their social environment depends on solidarity, togetherness, mutual understanding, mutual trust, clear communication and, depending on which (collective or social) goal to achieve, it requires collective action and effective cooperation (see Sect. 4.9). Society cannot rely on the market or state alone for solutions to collective problems, nor leave it to individual responsibility, so the power of collective action deserves special attention. Society could be reorganized to allow for more problem solving at the community level (the subsidiarity principle) and by forming new coalitions in horizontal innovation networks. This will require research into social-ecological interactions and interdependencies between stakeholders in complex change processes around nature, agriculture, land use, housing, mobility, and environmental issues and related policies (Aarts 2018). Research shows that 'people are good group problem solvers even if they are poor solitary truth seekers' (cf. Chambers 2018). Within this context, psychological and political science research shows that certain contexts motivate people to be reasonable, benevolent, and cooperative (Mendelberg 2002; Mercier and Landemore 2012; Chambers 2018; Dryzek et al. 2019). For example, small face-to-face group discussions can encourage individuals' cooperation (von Borgstede et al. 2013) and benevolent decisions (Mendelberg 2002). Within this context, people intend to engage more with environmental protection when they believe that future societies at risk of climate change will be more benevolent (Bain et al. 2013). People's reciprocity with their natural environment could be related to one's engagement in environmental issues

Table 3.3 Aspects of human nature that could be re-invigorated through a Natural Social Contract

Aspect of human nature	Explanation
Human life is group life	Humans flourish when living in families, and as a member of a group, community, collective, organisation, or company. Increasing levels of emotional, social, and psychological well-being result from participation in social relations (Keyes 2002; Fredrickson and Losada 2005; Snyder et al. 2011). To pursue self-preservation and higher levels of well-being, people depend on group life and reciprocity with their social and natural environments and related stewardship. The individualistic nature of capitalist societies conflicts with this important aspect of human nature. Instead, society could reorganize to nourish social cohesion, collective action, and problem solving at the community level while recognizing that some problems require the governance of social and political issues at the most appropriate level (i.e. the subsidiarity principle) and often requires multilevel governance. For example, the COVID-19 pandemic presents a societal challenge that requires more centralized and international coordination and crisis management.
Sharing	'Sharing' is an important evolutionary trait of humans. However, sharing has largely faded in capitalist societies and related economic systems where individualism, profit, and private property have superseded our relationship with each other and nature. In particular, the trait of sharing became skewed due to commercialisation of the public realm, rapid economic and technological change, and the rise of competitive individualism (Agyeman and McLaren 2017). In a Natural Social Contract, the human trait of sharing could be re-invigorated in the form of a sharing economy as well as in a circular and regenerative economy (see Sects. 3.3 and 3.4, and 'Economic dimension') and by co-management of the commons (see Sect. 4.3): natural or cultural resources available to all members of a group or society. This includes shared fishing waters, forests, agricultural land, and urban commons, as well as sources of information (e.g. Wikipedia), knowledge and culture.
Individual and collective learning	Humans possess a unique and unlimited ability to learn, gain new skills, and adapt to new environments and circumstances. Collective learning processes based on social relationships and networks are important for coping with uncertainty, enabling change, and developing the knowledge and ability to respond to new insights and developments (See Sect. 4.7 for more information on collective learning).
The power of imagination	The ability to form mental pictures or ideas in the minds of people is a powerful tool that supports intrinsic motivation, individual and collective learning, problem-solving, co-creation, and innovation. As Albert Einstein said, 'Imagination is more important than knowledge. For knowledge is limited, whereas imagination embraces the entire world, stimulating progress, giving birth to evolution'. See Sect. 4.11 for more information on co-creation.
Storytelling and narratives of change	Storytelling is the oldest way to share knowledge and ideas. From the oral tradition of ancient people to the written word and now in the digital age, it is stories that engage and compel us to understand new phenomena. When combining the strengths of stories—for instance about sustainability heroes—with that of system's thinking

(continued)

3.8 Dimensions and Crossovers Within a Natural Social Contract

Table 3.3 (continued)

Aspect of human nature	Explanation
	it provides a powerful approach for transformative learning (e.g. see Tyler and Swartz 2012), and the application of complexity thinking in all social-ecological systems. This combination of storytelling and systems thinking in order to facilitate transformative learning and institutional change is known in literature as 'Narratives of Change' (Krauß et al. 2018; Wittmayer et al. 2019).
The art of deliberate co-creation	Through the manifestation and sharing of ideas in social networks, humans command a unique trait for co-creating knowledge and advancing solutions. Co-creation involves collective problem solving with multiple parties and where multiple parties recognize mutual dependency and the importance of finding common ground, shared values, and mutually accepted solutions (see Sect. 4.11). Again, the power of imagination is important for co-creation when manifested in a tangible and joint vision based on a shared future. Such cooperation helps identify and create shared and common values during the co-creation trajectory, which in turn strengthens the bonds of actor coalitions (see Sect. 4.11).

Fig. 3.5 From ego awareness to eco-awareness (Edited figure, based on Scharmer and Kaufer 2013)

(e.g. through membership of a civil society organization or as an employee of an organization or company working on sustainability issues), or participation in the joint and sustainable management of public commons (see Sect. 4.4), or a deliberate decision for a sustainable lifestyle in one's household, for example, to opt for renewable energy, sustainable housing, sustainable food consumption, to limit one's carbon-based travelling, etc.

At the individual level, the sustainability transition is a period in which people start to look at their own lives, their lifestyles, and the consequences with a new set of eyes. This process is also known as a transition from ego awareness to eco-awareness (Scharmer and Kaufer 2013), as shown in Fig. 3.5. Similar to a global citizen or cosmopolitan—someone who feels involved in the world and is actively committed to making the world a better and safer place—the sustainability transition requires a rethink of one's citizenship, which could involve a behavioural change towards a more sustainable lifestyle or participating in collective action for sustainability. This could translate directly into one's voting behaviour or encourage more sustainable consumption of products (e.g. sustainable food, less meat, recycled materials, renewable energy) and services (e.g. shared or green mobility, circularity, green investments, support for local food producers and short food chains, or participation in community-supported agriculture). 'The role of residents might shift from receiving services and bearing rights to becoming more active in their immediate living environment and being subject to duties' (cf. Wittmayer et al. 2017). More in general, it involves a rethink of one's relation with other people and with nature, changing one's behaviour to increase the solidarity and reciprocity with one's social and natural environment and related stewardship. A popular course of action and 'leitmotiv' for global citizenship is 'think global, act local', while the sustainability transition is expected to bring about a growing number of 'Green Cosmopolitans', seeking options for low carbon travelling, more sustainable consumption, cleaner technologies, and collective action for sustainability. In addition, citizen science, in particular for biodiversity monitoring, is becoming increasingly popular (Sect. 4.10). 'As responsible citizens of planet earth, we can actively participate in the co-creation of actionable knowledge and solutions' (cf. Santha 2020).

The transition from ego- to eco-awareness and related behaviour is also known in literature as a transformational change in humanity from the '**Homo Economicus**' to the '**Homo Ecologicus**' (Dryzek 1996; Bosselmann 2004; Becker 2006; Cecchi 2013). The 'Homo Economicus' is defined as a rational person who pursues wealth for his own self-interest and was first mentioned by John Stuart Mill in the nineteenth century. In economic theory it is a model of human behaviour that assumes that people will make choices in their own self-interest. The assumption of rationality—also called the theory of rational behaviour—is primarily a simplification that economists make in order to create a useful model of human decision-making. Modern behavioural economists have disputed this theory, noting that human beings are actually irrational in their decision-making. Likewise, the concept of 'Homo Economicus' has received substantial criticism from other disciplines, either because of its misunderstanding of how social agents operate (Bourdieu 2005), and taking rationality of individual behaviour as the unquestioned starting point of economic analysis (Foley 1998), or because of the limited empirical outputs of rational choice theory (Green and Shapiro 1996). The 'Homo Ecologicus' provides a different model of human behaviour, which is closer to its natural behaviour, and provides a response to many of the above critics, in particular by zooming in on a human being's relationship (i) with itself, (ii) the community, and (iii) nature (Becker 2006). The *'Homo Ecologicus'* turns out to be inescapably social, unlike 'homo economicus' (cf. Dryzek 1996). In Henryk Skolimowski's ecological humanism, the concept is used to emphasize an equal position between humans and nature (Fios and Arivia 2018). As such, the concept of 'Homo Ecologicus' aligns with criticism on anthropocentrism by modern philosophers such as Bruno Latour, Henk Manschot, and Harry Kunneman, who argue for a new relationship with the Earth and other living beings.

> **Nature Connects**
>
> *"According to a series of field studies conducted by Kuo and Coley at the Human-Environment Research Lab, time spent in nature connects us to each other and the larger world. Another study at the University of Illinois suggests that residents in Chicago public housing who had trees and green space around their building reported knowing more people, having stronger feelings of unity with neighbors, being more concerned with helping and supporting each other, and having stronger feelings of belonging than tenants in buildings without trees. In addition to this greater sense of community, they had a reduced risk of street crime, lower levels of violence and aggression between domestic partners, and a better capacity to cope with life's demands, especially the stresses of living in poverty.*
>
> *This experience of connection may be explained by studies that used fMRI to measure brain activity. When participants viewed nature scenes, the parts of the brain associated with empathy and love lit up, but when they viewed urban scenes, the parts of the brain associated with fear and anxiety were activated. It appears as though nature inspires feelings that connect us to each other and our environment."*
>
> Text from: www.takingcharge.csh.umn.edu/how-does-nature-impact-our-well-being, retrieved on 1-10-2020.

Literature on positive psychology provides valuable insights on Human Flourishing, an approach aimed at increasing levels of emotional, social, and psychological well-being, due to participation in social relations (Keyes 2002; Fredrickson and Losada 2005; Snyder et al. 2011). The concept of Human Flourishing has many applications to civic duty and social engagement. Keyes (2002) shows that most people are more concerned with personal achievements than social relationships, but that this does not necessarily improve their well-being, so Keyes argues that children and adults should be encouraged to participate socially, because it improves the feeling of well-being and fulfilment. Could this be the case for humans in relation to nature as well? Throughout history, nature has had a leading role as a source of inspiration for musicians, visual artists, and scientists and will always be a driving force of creative inspiration. Being in nature, or even viewing scenes of nature, reduces anger, fear, and stress and increases pleasant feelings (Mitchell 2013; Russell et al. 2013; Sandifer et al. 2015) and could even reduce mortality (White et al. 2019). Various studies provide evidence for the positive impacts of human's interaction with nature (see text box 'Nature Connects' for an illustration).

Literature on social psychology provides valuable insights on norm interventions with regard to sustainable behaviour. For instance, Schwartz's norm-activation theory (Schwartz 1977; Schwartz and Howard 1981) is a model that explains altruistic and environmentally friendly behaviour. Other relevant social norm theories include Festinger's (1954) social comparison theory, Bandura's (1977) social learning theory (see Sect. 4.8), and Cialdini and colleagues' (1990) focus theory of normative conduct. An increasing number of studies is focusing on norm interventions with regard to sustainable behaviour (De Leeuw et al. 2015; Sparkman and Walton 2017; Thomas et al. 2017; Manomaivibool et al. 2016; Stöckli et al. 2018), including studies that integrate the Norm Activation Model (NAM) and Theory of Planned Behaviour (TBP) to understand sustainable behaviour, for example, in transport (Liu et al. 2017) and food sustainability (Onwezen et al. 2013). The theory of planned behaviour, with its origins in behavioural sciences, suggests that 'important actions are intentional and that the intention to act in a certain way is the immediate antecedent and cause of the behaviour' (cf. Ajzen 1991).

Social practice theory suggests that 'group behaviour is shaped by a combination of cultural norms and habits, rules and regulations, modes of provision, and infrastructures that together determine the ways in which people behave' (cf. Strengers and Maller 2014). This is particularly significant in the context of governing processes of change towards a sustainable society (ibid.). Norms and values that are central to a Natural Social Contract are discussed under 'Institutional Dimension' in this section. Within the social dimension of a Natural Social Contract, there are various aspects of human nature that could be re-invigorated (see Table 3.3), including storytelling and narratives of change. Storytelling is the oldest way to share knowledge and ideas. From the oral tradition of ancient people to the written word and now in the digital age, it is stories that engage and compel us to understand new phenomena. When combining the strengths of stories—for instance about sustainability heroes—with that of system's thinking it provides a powerful approach for transformative learning (e.g. see Tyler and Swartz 2012), and the application of complexity thinking in all social-ecological systems. This combination of storytelling and system's thinking in order to facilitate transformative learning and institutional change is known in literature as 'Narratives of Change' (Krauß et al. 2018; Wittmayer et al. 2019).

Beyond the individual or citizen level, a transformation towards a Natural Social Contract assumes a change in behaviour and shifting roles for all societal actors. A social contract is an empty shell without participation in all layers of society. It requires a rethink of how each actor could contribute to the sustainability transition. For example, traditional banks as an institution are changing towards banking as a service adjusted to digital innovation, decentralized and sustainable production and consumption (Ryszawska 2018). Pension funds, with USD 28 trillion in assets (OECD 2011), along with other institutional investors, have an important role to play in financing the sustainability transition. Businesses have a decisive role to play in the transition from linear to circular business models. Civil society could play a number of roles in sustainability transitions beyond civil advocacy (Frantzeskaki et al. 2016), while 'governments are searching for different relationships between governments, institutions, and citizens, from active financial commitment linked with targets, and moving from controlling and containing to facilitating and supporting' (cf. Wittmayer et al. 2017). Higher education must prepare students for their future and educate them to think across sectoral boundaries and favour transdisciplinary approaches, which will require new knowledge and skills. This will address the grand societal challenges such as climate change mitigation and adaptation, circularity, urban health, citizen participation, food transition, and the energy transition (Sect. 7.5).

Ecological Dimension

A Natural Social Contract explicitly emphasizes the natural position of humankind and society within a larger ecosystem, that of planet Earth. All life on earth is interconnected, interdependent, and subject to the same set of circumstances (e.g. sunlight, water, gravity, cyclical processes, complex adaptive systems, and non-linear feedback loops). Earth is the whole, within which humans are subservient (though impactful) parts. Hence, in a Natural Social Contract, society is viewed as a social-ecological system (SES), in which an ecological system is intricately linked with and affected by one or more social systems. In the past two decades, the concept of social-ecological systems (SES) has become central to an increasingly widespread international discourse on human–nature interactions (Berkes et al. 2000; Anderies et al. 2004; Olsson and Galaz 2012; Becker 2012; Skandrani 2016; Rissman and Gillon 2017, and others). Olsson and Galaz (2012) argue that only addressing the social dimension will not be sufficient to guide society towards sustainable outcomes. A societal transformation towards sustainability requires improving society's capacity to learn from, respond to, and manage environmental feedback from dynamic ecosystems (ibid.). For example, 'a systemic shift to biofuels might slow climate change, but lead to destructive land use change and biodiversity loss (Grau and Aide 2008). This in turn can lead to further ecological degradation, regime shifts, and lock-in traps in social-ecological systems that are difficult to get out of' (cf. Olsson and Galaz 2012).

Within the context of the sustainability transition, design lessons taught by nature, such as adaptive capacity, resilience, resource efficiency, circularity, self-organization, and the networked relationship between all organisms, deserve special attention (see Sect. 4.4). The similarities between well-functioning social and ecological systems are very large, but people have lost sight of some basic principles. In a healthy, mature ecosystem nothing is wasted, with full circularity of energy and matter, just to mention one thing. Could this be an example for a wasteless and circular society? The Cradle to Cradle movement and the ongoing transition to circular and regenerative economies and cultures make grateful use of these insights from ecology. Redesigned economics based on ecology, such as in Circular Economy and Regenerative Economy, will play a central role in developing a Natural Social Contract, at least from an economic perspective (see Sect. 3.3).

> "A watershed includes all the humans, plants and animals who live in it, and all the things we have added to it such as buildings and roads. Everything we do affects our watershed—from washing clothes and growing food to mining, commercial farming, and building roads or dams. The reverse is also true: our watershed affects everything we do, by determining what kinds of plants we can grow, the number and kinds of animals that live there, and how many people and livestock can be sustainably supported by the land. One important

(continued)

> truth about watersheds is that we all live downstream from someone, and upstream from someone else. Anything dumped on the ground in the watershed can end up in its rivers, lakes or wetlands. And anything released to the air can come down again, nearby or thousands of miles away. We are all connected through watersheds. Watersheds do not respect political or administrative boundaries, and in fact can encompass several cultural, national and economic boundaries."
>
> *Text from www.internationalrivers.org, retrieved on 13-4-2020.*

Valuable lessons can be drawn from societal transformations where insights from nature and principles of nature (see Sect. 4.4) have already led to paradigm shifts towards sustainability. For example, the water management regimes in low-lying countries such as the Netherlands and Vietnam have seen a paradigm shift in the past decades from the 'fight against water' towards 'living with water', by translating key lessons from ecology, in particular from resilience theory and adaptive management, into spatial planning for river basin management. For instance, the authorities have substantially improved the buffering capacity and resilience for peak discharges in the Rhine and Mekong river basins by undoing land reclamation and giving it back to nature and reconnecting the main rivers with wetlands, creating green bypasses, and broadening of floodplains by dyke replacements (Huntjens et al. 2011a, b, 2018). This paradigm shift in water management became possible once planners and engineers started to look at the rivers from an ecological point of view, in which land and water are linked in a natural system called a catchment, drainage basin, or watershed (see text box). The insight that a river basin can only become resilient and healthy through a basin management approach, and in case of the Rhine river basin and Mekong rivers basins thus requiring international cooperation, prompted the establishment of the Rhine and Mekong River Commissions and served as examples for European water policies and river basin management worldwide ever since (Heldt et al. 2017; Van Diep et al. 2007). Such a river basin management (RBM) approach is not a blueprint though and requires adaptation to the different socio-economic, cultural, political, and biophysical contexts of the implementing countries.

The above examples explain why the concept of biomimicry (Sect. 4.4) has increased in popularity, as has, more recently, the interest in biomimicry in the context of social innovation, with the aim of creating products, processes and policies that are well-adapted to life on earth in the long term (Benyus 1997; Biomimicry 3.8 2013; Fullerton 2015; Wahl 2016). Within a Natural Social Contract, biomimicry should not be taken as an 'imitation' of life as much as a 'return' to natural, sustainable behaviour by humankind as a component of a greater ecosystem, that of planet earth. However sad this observation may be, the concept of biomimicry alone confirms that humankind lost its way at some point and stopped seeing itself as something 'natural', to the point that we are now forced to mimic nature for the purposes of self-preservation, increasing well-being and to safeguard the liveability of the planet for future generations. That is why I would argue that a Natural Social Contract, and the related process of Transformative Social-Ecological Innovation (TSEI), does not constitute mimicking nature, but rather constitutes a return to our origins. In essence, it means moving forward by looking backward. A Natural Social Contract determines the structure and functioning of a sustainable, healthy, and just society, in particular based on design lessons taught by nature. When combining these insights from nature and principles of nature (see Sect. 4.4) with the lessons learned from modernization and civilization processes and social contract theory, in particular humanity's quest for security and justice for all, it provides a logical and powerful combination for establishing a Natural Social Contract.

Economic Dimension

From an economic perspective, a Natural Social Contract reserves a central place for Circular, Regenerative, and Sharing/Collaborative Economies, which are forms of alternative economies that closely resemble natural human behaviour (see social dimension above, in particular on 'Homo Ecologicus') and natural design principles (Sect. 4.4 on biomimicry).

A Sharing/Collaborative Economy departs from the human evolutionary trait of sharing, which has largely broken down in capitalist societies (Agyeman and McLaren 2017), mainly due to the commercialization of the public realm, rapid economic and technological change, and the rise of competitive individualism (ibid.). In a Sharing or Collaborative Economy 'access' to goods and services is more important than 'ownership' of goods (Belk 2014; Barbu et al. 2018). A common premise is that when information about goods is shared (typically via an online marketplace), the value of those goods may increase for the business, for individuals, for the community, and for society in general. In this vein, Mason (2016) argues for more cooperative schemes of free exchange—a 'sharing' economy to replace a predatory one—and more collective ownership as well. For example, carsharing is part of a larger trend of shared mobility, which is different from car rental in that the owners of the cars are often private individuals themselves, and the carsharing facilitator is generally distinct from the car owner. In particular, the transition to a Sharing or Collaborative Economy requires a change from traditional market behaviour to collaborative consumption models, in which resources are used more efficiently and sustainably. This economic model is therefore part of a broader transition to a circular and regenerative economy and offers business models that are compatible with it (Barbu et al. 2018).

Similar to the alignment with natural human behaviour, as in the case of a Sharing or Collaborative Economy, a Natural Social Contract underscores the importance of economic design based on, or inspired by, lessons taught by nature and natural design principles (see Sect. 4.4). In particular design lessons taught by healthy and mature ecosystems deserve special attention, such as adaptive capacity and resilience, resource efficiency, and circularity. Circular Economy and Regenerative Economy are examples of economic design based on ecology, where nature shows how circularity is usually organized at the lowest possible level. Likewise, circularity needs to be understood as a property of a system, such as the mobility system of a city, rather than a property of an individual product or service, for instance, a car or a carsharing service (Konietzko et al. 2020). These new economic models promise to organize different forms of sustainability at different levels, where the prevention of waste and the valuation and revaluation of all matter and resources are important starting points. A large part of our clothing, furniture, electronics, and our food ends up at a dumping ground for waste or is incinerated, while it still contains a lot of valuable raw materials. Much more can be done, but getting there requires technological and social innovation, including organizational and institutional innovation. The ratio between the impact of technology and social innovation for realizing a circular economy is estimated at 25:75 (Jonker et al. 2018).

Within the school of Ecological Economics, which originated in the 1980s, the economy is treated as a subsystem of Earth's larger ecosystem and addresses the interdependence and coevolution of human economies and natural ecosystems, both intertemporally and spatially (Xepapadeas 2008; Van den Bergh 2001). The literature on Ecological Economics emphasizes that the natural world has a limited carrying capacity and that its resources are running out. Ecological economists assume 'that growth is *not* a given, and that population growth, inequalities, and the decline of cheap and abundant fossil fuels, which spurred the unprecedented growth of the global economy over the past century, mean that the limits to growth are either being reached or will be reached in the very near future' (cf. Caradonna et al. 2015). 'Since the destruction of important environmental resources could be catastrophic and practically irreversible, ecological economists are inclined to justify cautionary measures based on the precautionary principle' (cf. Costanza 1989).

By contrast, 'Ecomodernism' is a school of thought 'that emphasizes the roles of technology and economic growth in meeting the world's social, economic, and ecological challenges' (cf. Caradonna et al. 2015). The *Ecomodernist Manifesto* (2015) rejects the idea 'that human societies must harmonize with nature to avoid economic and ecological collapse. Instead, it argues for more reliance on technologies, from nuclear power to carbon capture and storage. Many ecomodernists ridicule the idea of limits to growth, arguing that technology will always find a way to overcome those limits. For instance, they believe we can feed the world with more intensive agriculture, the combination of hybrid seeds, high-intensity fertilizers, precision agriculture, and making crops more productive with *genetically modified organisms* (GMOs). Basically, keep doing what we have been doing but make it cleaner' (ibid.).

Conversely, ecological economists consider industrial agriculture as the problem causing environmental degradation. For energy supply, 'they propose renewable energy sources (e.g. wind and solar), instead of nuclear power, since the latter will lead to long-term storage nightmares and present-day environmental hazards (e.g. see Chernobyl and Fukushima). For carbon capture they propose planting trees instead of technical solutions. For feeding the world, they would like to see a radically different form of agriculture (e.g. based on organic agriculture, permaculture, and food forests, which also reduce the output of carbon), one that does not depopulate the countryside, one that mimics natural ecosystems and grows lots of types of crops' (cf. Johnson 2018). It requires more work than industrial solutions, but would create more jobs and vital rural communities. 'Ecological economics shares several of its perspectives with feminist economics, including the focus on sustainability, nature, justice, and care values' (cf. Aslaksen et al. 2014).

While Ecomodernists value individual liberty more, Ecological Economists place more value on communities, but there is no law of physics that says you cannot have both (Mann 2018). There is an obvious middle ground, and more attention should be devoted to a mix of solutions and to bridge the divide between both camps. For example, for developing climate-resilient coastal areas and cities (see Chap. 7), it makes perfect sense to use a portfolio approach of desalination technologies, breeding of drought or salt-tolerant species, renewable energy, and other technological solutions, in combination with reforestation, circular and regenerative agriculture, and community-based approaches, thus combining the best of both worlds. For a regional and circular food system (see Chap. 6), it makes sense to combine both high-tech methods, such as vertical farming, robotization, artificial intelligence and information technologies, and low-tech methods, such as vegetable gardens, urban gardens, food forests, and regenerative agriculture. With a shared vision, part of these (future) production systems is high-tech, while the systems are in harmony with nature, with a fully transparent and zero-emission food chain, including food producers and consumers as co-owners of their own food system. Recently, interest in local food systems has skyrocketed (Pigford et al. 2018), centred around an approach in which citizens, farmers, and other stakeholders work together to create a sustainable, healthy, and predominantly local food system (see Chap. 6 for examples).

The majority of products and services in our current market-based economies leads to increasingly higher and hidden societal and ecological costs. Think of environmental and climate damage, damage to human and animal health, and underpayment of farmers and health workers, for instance. These costs are not included in the market price of the product or service. Making these costs visible and including these costs in the prices of products and services is an important avenue for creating a level playing field with more sustainable, healthy, and fair products and services. This is the essence of True Cost Accounting (TCA) or Full Cost Accounting (FCA), which represents a rapidly growing academic discipline (Negowetti 2016; Aspenson 2020), in particular in the food and clothing industry, focused on calculating the impact that products and services have on natural, human, and social capital—the so-called business externalities. Usually these impacts—positive or negative—are not reflected in the prices paid by users/consumers. Hence, calculating true costs, based on the discounting of integral costs such as CO_2, nitrogen, toxicity, or living wage, is an essential step in creating an economic model in which transparency, sustainability, health, and fair prices and incomes are central rather than high volume, low prices, and low incomes. With the vast majority of consumers usually opting to pay the lowest price, it prompts the food and clothing industry to adopt highly efficient, low-lost production methods. As a consequence, there is little incentive for actors in these production chains to invest in sustainability measures and translate those into cost price. This economic logic leads to a vicious circle and a race to the bottom. Hence, True Cost Accounting is an important intervention to break this vicious circle. In the coming decade, the incorporation of true prices into our economic systems will be an important systemic change towards a Natural Social Contact.

Institutional Dimension

By taking Calhoun's definition (2002) of institutions as 'deeply rooted patterns of social practices or norms that play an important role in how society is organized' it becomes evident that the institutional dimension has close ties with every dimension of a Natural Social Contract. A further distinction is made between formal institutions (those adopted through a formalized process, including the constitution, laws, and legislation) and informal institutions (those embedded in organizations or groups without a formalized process, including customary law, existing practices, norms, and culture).

There are relevant crossovers between the ecological dimension and the institutional dimension within a Natural Social Contract. Examples of institutional design based on ecology include governance approaches that are better capable of dealing with complexity and uncertainty, such as adaptive, deliberative, and reflexive governance (Sect. 4.7). Rather than considering complexity and uncertainty as difficult obstacles that must be controlled, mitigated, or ignored, these characteristics of social-ecological systems should be considered drivers of Transformative Social-Ecological Innovation (TSEI). Other examples of ecological-institutional crossovers include the precautionary principle and polluter pays principle as important mechanisms for steering social-ecological systems towards sustainability. Furthermore, insights from the governance of the commons and the sustainable co-management of natural resources (e.g. fishing grounds, forests, and agricultural land) and cultural resources (e.g. sources of information, knowledge, and culture) are relevant for the institutional design of a Natural Social Contract (see Sect. 4.3).

A Natural Social Contract necessitates governance at a level of scale that does the most justice to the complexity of socio-ecological systems, for example, through polycentric governance (Sects. 4.3 and 4.6). Within this context, the principle of subsidiarity, one of the core principles of European Law, prescribes the governance of social and political issues at the most appropriate level. Section 3.10 provides examples for the development and implementation of a Natural Social Contract at various governance levels, ranging from the local to the national and international level. Within this context, adaptive governance of social-ecological systems generally involves polycentric institutional arrangements (see Sect. 4.3), 'which are nested quasi-autonomous decision-making units operating at multiple scales (Ostrom 1996; McGinnis 2000). They involve both local and higher organizational levels and aim to find a balance between decentralized and centralized control (Imperial 1999). The term multi-level governance is used to characterize the relationship between actors situated at different administrative and territorial levels. This creates layers of actors who interact with each other: (1) across different levels of government (vertical coordination); (2) among relevant actors at the same level (horizontal coordination at central or at subnational level); or (3) in a networked manner. This relationship exists regardless of the constitutional system (federal or unitary) and impacts the implementation of public policy responsibilities. Debates over "scaling" powers within multi-level governance have become widely discussed in several related academic sub-disciplines, including economic federalism (e.g. Oates 1998), political geography (e.g. Delaney and Leitner 1997), EU studies (Hooghe and Marks 2003; Bache et al. 2016; Hooghe et al. 2020), and international public policy (Young 2002). For example, conflicts over the appropriate "scale" (Young 2002) or institutional level of policymaking characterize multi-level governance' (cf. Huntjens 2011). In the past two decades, multi-level governance has become an important concept in climate change and environmental policies (e.g. see Di Gregorio et al. 2019; Hooghe et al. 2020) and is often used to capture the dynamics of EU cohesion policy (Bache et al. 2016).

Finally, on the topic of governance, many of the institutional arrangements relevant for a Natural Social Contract are reflected in Good Governance Principles, developed and adopted by the United Nations (UNESCAP 2009) and the Council of Europe (CoE 2008) among others, covering issues such as ethical conduct, rule of law, efficiency and effectiveness, transparency, sound financial management, and accountability. These principles are applicable to corporate, international, national, or local governance. It should be clear that good governance is an ideal which is difficult to achieve in its totality. Very few countries and societies have come close to achieving good governance in its totality (UNESCAP 2009).

Between the economic dimension and institutional dimension there is a variety of crossovers. As already mentioned under economic dimension, for realizing a circular economy it requires technological and social innovation (including organizational and institutional innovation) with a ratio of 25:75 (Jonker et al. 2018). Likewise, a number of institutional innovations have already been tested for realizing a sustainable, healthy, and predominantly local food system (see Chap. 6), such as Food Policy Councils (FPCs), as loci for practising food democracy (Sieveking 2019; Scherb et al. 2012; Gupta et al. 2018; Sussman and Bassarab 2017), community-supported agriculture (CSA), as a sustainable alternative for industrial agriculture (Kondoh 2015; White 2020; Galt 2013), and Short food supply chains (SFSC), which aim 'to reconnect the two extremities of the food supply chain, reconcile producers with citizens, stimulate mutual trust, and establish a short chain based on common values on food, its origin and production method' (cf. SKIN 2020).

Taxation is a powerful tool for steering the behaviour of both producers and consumers. Taxation policy thus relates directly to a Natural Social Contract's economic and social dimension as described in this section, but certainly also to the institutional dimension. Several recent studies show that sustainability-oriented taxation is an effective tool for mitigating unsustainable and unhealthy behaviour, products and services, for instance, through carbon taxes (Krenek and Schratzenstaller 2016; Ulucak and Kassouri 2020). On top of that, fuel taxation may be a promising public health intervention for obesity prevention (Brown et al.

2017), since it would become more expensive to use motorized vehicles, and instead encourages people to walk or take a bicycle more frequently. Tax revenues, in return, can be used to provide positive incentives for sustainable and healthy practices, behaviour, products, and services while helping to offset the costs of the sustainability transition, in a socially just manner. Likewise, taxation policy could play an important role in addressing growing inequality. Empirical studies show that inequality is increasing (Piketty 2013; Kremer and Maskin 2006). When labour is more heavily taxed than wealth, the rich are getting richer, while the poorer part of the population succumb to make ends meet (Scheve and Stasavage 2016). The tax on capital must therefore increase considerably, that on labour considerably down. Higher taxes from the rich would pay for programmes that improve the welfare of the poor through the government's expenditure policy. Hence, given its role in feeding budgets, distributing resources, and steering behaviours, taxation has a pivotal role in a societal transformation towards a Natural Social Contract.

Not surprisingly, there are close ties between the social and institutional dimensions in a Natural Social Contract, of which perhaps the most prominent include norms and values, which are either adopted in formal institutions or embedded in informal institutions.

Core values of a Natural Social Contract should be made explicit and discussed in any process of Transformative Social-Ecological Innovation and is often a necessary step in the process of creating shared value and multiple value creation (Sect. 4.8). The focus theory of normative conduct proposes, for example, 'that norms are important to the extent they are made salient at the time of action, such that individuals make behavioural decisions on the basis of normative considerations, rather than other considerations' (cf. Cialdini et al. 1990).

Common values in a Natural Social Contract could be context-specific to some extent, tuned to specific features of local geography, ecology, economies, and cultures, but also include a certain level of universality, especially when reasoning from a human being's natural state as a social animal living in families and thus requiring some level of collectivity, solidarity, mutual trust, and reciprocity. Chances of survival are larger when operating as a group, requiring clear communication and effective cooperation. Common values also appear when looking at general and historical patterns of civilization, modernization, and the human need for social order, security, and justice. Schwartz and Bilsky (1987) defined 'values' as 'conceptions of the desirable that influence the way people select action and evaluate events'. They hypothesized that 'universal values would relate to three different types of human need: biological needs, social coordination needs, and needs related to the welfare and survival of groups' (ibid.). The claim for universal values can be understood in two different ways. First, it could be that something has a universal value when everybody finds it valuable. Second, something could have universal value when all people have reason to believe it has value' (cf. Jahanbegloo 1991). 'When Mahatma Gandhi argued that non-violence is a universal value, he was arguing that all people have reason to value non-violence, not that all people *currently* value non-violence' (cf. Amartya Sen 1999, page 12). The same reasoning could be applied to values related to a Natural Social Contract, in which all people have reason to believe that solidarity and togetherness (as being central to group life), environmental protection, and the sustainable and joint management of shared resources, collective well-being, democracy, and justice are valuable and thus accepted as general principles for governing day to day life.

In the past few decades, values on the relationships between humans and nature are becoming more prominent and recognized in initiatives, policies, and laws around the world (Sect. 3.10). In particular, the Rights of Nature as a legal and jurisprudential theory, which describes inherent rights associated with ecosystems and species, deserves special attention. While twentieth and twenty-first century environmental laws do afford some level of protection to ecosystems and species, it is argued that such protections fail to stop, let alone reverse, overall environmental decline, because nature is by definition subordinated to anthropogenic and economic interests, rather than the well-being of non-humans and nature (Cullinan 2011; Berry 1999; Biggs et al. 2017; Borràs 2016). Thomas Berry (2006) proposed that society's laws should derive from the laws of nature, explaining that 'the universe is a communion of subjects, not a collection of objects'. Just as human rights are increasingly being recognized in law, advocates claim that nature's rights must also be recognized and incorporated into human ethics and laws. This claim is substantiated by the same ethic that justifies human rights, and that the survival of human beings depends on healthy ecosystems (Cullinan 2011; Berry 1999; Stone 1996; Nash 1989). An obvious challenge to the Rights of Nature is that neither Nature in general, nor particular species and ecosystems have the kind of agency required to exercise and defend their rights. Environmental law scholars therefore suggest appointing a custodian to represent the Rights of Nature while taking precautionary measures to avoid an overly anthropogenic representation by such a custodian. While not without obstacles, the inclusion of ecocentric theories in legal frameworks is an important avenue for a societal transformation towards a Natural Social Contract.

The Gallup Institute's World Values Survey indicates an ongoing paradigm shift towards eco-awareness and post-materialistic value sets. The rise of postmaterialist values is part of a broader set of cultural changes that tend to bring democratization (Inglehart 2017) and a transition from ego awareness to eco-awareness (Scharmer and Kaufer 2013). This trend is reflected in public opinion, policies, and laws around the world, including the following examples:

- The United Nations World Charter for Nature, adopted in 1982, announced five principles of conservation by which all human conduct affecting nature is to be guided and judged.
- The Earth Charter—an ethical framework for sustainable development published in 2000—reserves a central place for environmental protection, human rights, equitable human development, and peace and argues that these values are interdependent and indivisible.
- Several countries have recognized Rights of Nature in their legal frameworks and/or jurisprudence, e.g. in Uganda, Peru, Ecuador, Mexico, Colombia, India, Bangladesh, New Zealand, and communities in the USA.

- Granting rights to rivers, e.g. the Whanganui River in New Zealand, the Yarra River in Australia, and the Ganges and Yamuna rivers in India.
- The Universal Declaration of Rights of Mother Earth, adopted by the World People's Conference on Climate Change and the Rights of Mother Earth, April 22, 2010 in Cochabamba, Bolivia.
- Intergenerational justice and equity, a range of immediate and effective mechanisms is proposed to safeguard the rights of future generations and protect them from the potential negative implications and harm caused by climate change and environmental degradation (Sect. 2.3).
- In many countries, public interest in animal welfare, animal rights, and plant-based diets has increased significantly (Grunert et al. 2018). The World Animal Protection (WAP) charity has successfully lobbied the United Nations to include language on animal welfare in two General Assembly Resolutions on agriculture and disaster risk reduction in 2013, but surprisingly, there is no mention of animal welfare in the UN Sustainable Development Goals adopted in 2015 (Visseren-Hamakers 2020).

Generally speaking, a Natural Social Contract reserves a central place for core values such as solidarity, togetherness, collective well-being (as being central to group life), democracy, equity, social and environmental justice, and social and environmental stewardship. The latter entails stewardship for, and reciprocity with, our social and natural environment, for example, through the sustainable co-management of natural resources (e.g. fishing grounds, forests, and agricultural land) and cultural resources (e.g. sources of information, knowledge, and culture). More specifically, a Natural Social Contract stresses the importance of values such as social and environmental stewardship. After all, everyone is part of a social and natural environment, and the environment is part of each of us. It is worth noting that values such as stewardship and solidarity have a prominent role in all world religions. For instance, many religions and denominations have various degrees of support for environmental stewardship, which is a theological belief that humans are responsible for taking care of the world, including all life (humans, animals, and nature). Another example comes from New Zealand, where the Maori term Kaitiaki is used for the concept of guardianship, for the sky, the sea, and the land. This concept has been adopted in New Zealand's legislation, allowing Maori communities to be appointed as guardians for a specific area.

The overall goal of a Natural Social Contract is to promote human and environmental security, social and environmental justice, and planetary health. This could be translated in a tangible vision of a sustainable, healthy, and just society where prosperity is broadly defined and fairly distributed, including the economic, ecological, and social dimensions of prosperity and sustainability, and with interventions designed to mitigate poverty, inequality, social exclusion, and environmental degradation. This vision must include gender equality, and interventions to ensure that women have the same prospects and opportunities as men, and interventions to protect the sick, the vulnerable, and minorities of all kinds. Such a tangible vision could serve as a vehicle to identify and create shared and common values during the process of Transformative Social-Ecological Innovation (TSEI). Agreement on these ethical and normative aspects is important for holding actor coalitions together during a transition process and could be achieved through deliberation on shared beliefs and values, shared discourses, common interests (Sect. 4.8), procedural justice (Sect. 4.9), and options for multiple value creation and mutual gains (Sect. 4.8).

Table 3.4 provides a summarized overview of this section in the form of design principles for a Natural Social Contract. This overview is shaped as a course of action and is intended to help readers to capture the core rationale of this book.

3.9 TSEI-Framework for Understanding and Advancing the Process Towards a Natural Social Contract

Social contract thinkers ask themselves how social and political order in society can be legitimized (Gabriels 2018). Although their opinions differ, Gabriels argues they all take the same three steps: 'First of all, they outline a baseline situation, a conflict situation that must be resolved by means of a contract. Secondly, contract thinkers present a procedure for agreeing on the content of the contract, which should offer various different solutions to a conflict situation. Thirdly, contract thinkers describe the results of the chosen procedure, i.e. the actual implementation of the contract'. In this section I introduce and propose a conceptual framework for Transformative Social-Ecological Innovation (see Fig. 3.6), with the purpose of providing a better understanding and advancing the process of developing and implementing a Natural Social Contract. In section I will follow the same three steps as described above by Gabriels (2018).

The **baseline situation** for a Natural Social Contract is outlined in Chaps. 1, 2, 3 and constitutes a complex set of security and justice problems that need to be resolved. In the previous chapters it becomes clear how the most comprehensive societal fault lines of our times are deeply intertwined and confronts us with challenges concerning the security as well as justice of our societies. Increasing wealth inequality, financial crises, ecological crisis, climate crisis, trade wars, migration issues, and even the Corona pandemic can be traced back to two common denominators. First, the schism between humans and nature, and the dominant anthropocentric world view that arose during the Enlightenment. Second, the capitalist economic logic, in particular the unsustainability of infinite economic growth in a finite world, and the belief in the infallibility of the free market, that arose after the Second World War. It left us with market-based societies that reserve a central place for individualism, materialism, privatization, short-termism, the free market, and a singular focus on profit and economic growth. In a market-based society citizens have been turned into customers and consumers and demand what they have paid for, but feel no duty to the community or environment. This led to decades of excessive production, consumption, and depletion of our natural resources and raw materials. The resulting loss of biodiversity and key ecosystem functions, environmental degradation, bio-industry, land, water, and air pollution, and fossil energy

Table 3.4 Fundamentals and design principles for (a societal transformation towards) a Natural Social Contract

Dimension	Fundamentals and design principles for a Natural Social Contract
Social	**Rediscover or reinvigorate community**: Human life is group life, and people are woven into a web of dependencies, and are generally happier if they can take care of each other and nature and know that they are being taken care of. Society could be re-organized in such a way that problems can be solved at the most appropriate level (the subsidiarity principle), while citizens should be involved as much as possible in decisions about their own living environment
	From 'Homo Economicus' to 'Homo Ecologicus': Aspires to interconnect individual and community with social and natural environments » Move from ego- to eco-awareness and beyond individualism to **social and environmental stewardship and solidarity**
	Human flourishing in a responsible way: Aims to increase levels of emotional, social, physical, and psychological well-being through our connections to and participation in our social and natural environments
	Encourage and support **collective and adaptive learning** processes » Advance information management through participatory knowledge creation, transdisciplinary research, and a commitment to dealing with uncertainties, as well as reflexive monitoring, broad communication between stakeholders, open and shared information sources, and flexibility and openness to experimentation (e.g. in living labs)
	Shaping **group behaviour and social-ecological interactions**: Governing processes of change towards a sustainable, healthy, and just society. This takes into account a combination of cultural norms and habits, rules and regulations, modes of provision, and infrastructures. See also 'Taxation' below
Economic	**Economy for human and planetary well-being, not for profit**: This requires us to broadly define welfare, including the economic, ecological, and social aspects of prosperity
	From linear economies (i.e. produce, use and dispose) towards local, circular, and regenerative economies and cultures: Designing economic models based on lessons from nature (e.g. circularity at the lowest possible level, short supply chains, local self-sufficiency in water, food, energy). This requires a transition from linear to circular business models characterized by collective and shared value creation
	Joint management of the commons instead of private ownership: Ensure sustainable co-management of natural resources (e.g. fishing grounds, forests, and agricultural land), urban commons, and cultural resources (e.g. sources of information, knowledge, and culture)
	Sharing economy: Facilitating shared access to goods and services (e.g. through a community-based online platform) to improve efficiency, sustainability, and community values
	Diversify financial resources: Apply a broad set of public and private financial instruments. This could include innovative and hybrid forms of banking and financing (e.g. revolving energy and sustainability funds)
	True cost pricing: Making visible the hidden social and ecological costs in the price of products and services. This encourages the creation of a level playing field with more sustainable, healthy, and fair products and services
Ecological	**Navigating complexity & embracing systems thinking**: Systems thinking sees life in continuous motion and recognizes that the larger picture is rarely static but rather almost always a web of factors that interact to create patterns and change

(continued)

Table 3.4 (continued)

Dimension	Fundamentals and design principles for a Natural Social Contract
	over time. Adopting a systems-based approach helps identify synergies and trade-offs that move beyond linear to more circular and inclusive systems
	Governing society as a social-ecological system: Earth is the whole in which humans are subservient (but impactful) actors. We must design new ways to inhabit and cultivate our planet to keep it healthy for future generations (e.g. through renewable energy, sustainable agriculture and water management, circular and regenerative economies, environmental protection, polluter pays principle, precautionary principle)
	Accept nature as a teacher, not as an enemy: Learn from, respond to, and manage environmental feedback from dynamic ecosystems. Economic and institutional design based on lessons learned from nature (in particular sustainable, healthy, and mature ecosystems). This includes adaptive capacity, resilience, resource efficiency, circularity, self-organization, and the interconnected relationship between all organisms
	Work with Mother Nature, not against Her: Urban and rural landscapes where ecology and economy encourage equilibrium that produce and protect at the same time (see examples of eco-cities, sponge cities, and circular and nature-inclusive agriculture in this book)
Institutional	**Invest in inclusive and deliberative local democracy and polycentric governance**: (1) Strengthening inclusive procedures to broaden legitimacy of decisions and actions, through stakeholder participation and involving all layers of society; (2) deliberating shared beliefs and values, common interests, procedural justice, and opportunities for multiple value creation and mutual gains; (3) encouraging consensus and collective action based on reasonable and evidence-based arguments where persuasion emphasizes mutual understanding and compromise and supports a process that is inclusive, open, trusting, and collective
	Adaptive, reflexive, and deliberative approaches to governance: Taking ambivalence, complexity, uncertainty, and distributed power into consideration in societal change. Governance, planning, and management at a level of scale that does the most justice to the complexity of social-ecological systems (subsidiarity principle)
	Security and justice for all: Promote the equal and fair distribution of wealth instead of rising inequality, including equal and fair (re-)distribution of risks, costs, and benefits through the involvement and strengthening representation of marginalized and particularly vulnerable groups and stakeholders. Ensure that women have the same prospects and opportunities as men; and protect the sick, the vulnerable, and minorities of all kinds
	Rule of Law and accountability: Recognize and incorporate **human rights** and the **Rights of Nature** into human ethics and laws. Adhere to the rule of law with regard to human and environmental security and justice while preventing and resolving conflicts through a variety of mechanisms (e.g. expanding access to justice and dispute resolution mechanisms)
	Taxation: Mitigate unsustainable and unhealthy behaviour, products, and services through effective taxation policy (e.g. carbon tax). Tax revenues and tax rebates can offer positive incentives for sustainable and healthy practices, behaviour, products, and services. Additionally, taxation addresses **growing inequality** by enacting progressive tax policies on capital and lowering tax burdens on labour

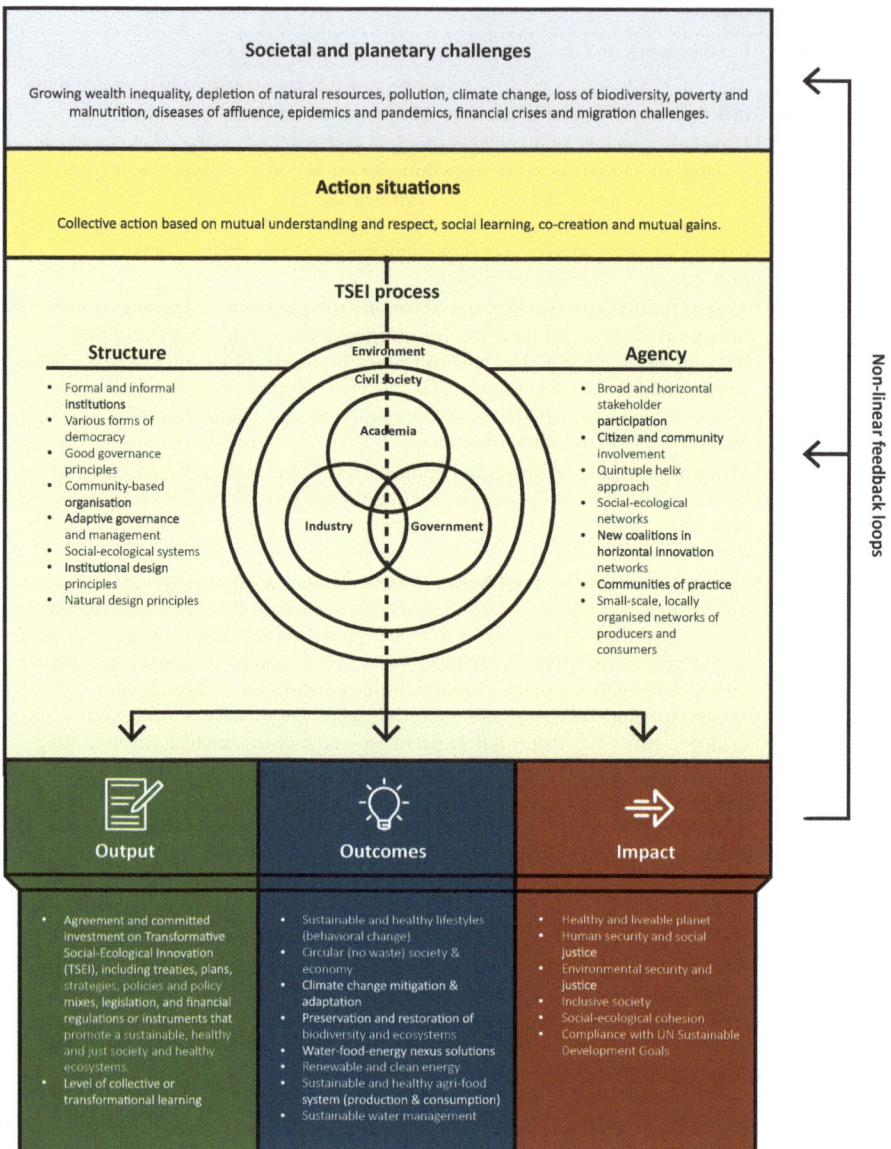

Fig. 3.6 Conceptual framework for Transformative Social-Ecological Innovation (TSEI)

consumption, for instance, show that the way we deal with nature is profoundly disturbed. We are now discovering that ecological vulnerability translates into economic and social vulnerability and a complex set of security and justice challenges.

The **procedure** describes how an agreement about the contents of a social contract can be achieved. A social contract is not a formal contract of course, but something that is lived and practised. It is something that has to grow, but the mechanisms of its growth and interaction with existing orders are not well understood. The conceptual framework for Transformative Social-Ecological Innovation (TSEI) presented below is developed with the objective to provide a better understanding of the 'procedure' for realizing a Natural Social Contract (see Fig. 3.6, with more detail in Chaps. 4 and 5). In this book I emphatically replace the more formal term 'procedure' with 'process', since TSEI does not specify a blueprint, but encourages transformative processes tuned to the specific features of local geography, ecology, economies, and cultures.

The Transformative Social-Ecological Innovation (TSEI) framework presented in this book offers new ideas for unpacking and understanding institutional change across sectors and disciplines and at different levels of governance. To this end, it identifies intervention points and helps to formulate sustainable solutions that can include different views, as well as changing and competing needs. Overall, the concept of TSEI encourages public officials, business leaders, and the greater public to think more broadly about how society can rethink cooperation to address humankind's greatest challenges.

Transformative Social-Ecological Innovation (TSEI) is inextricably tied up with questions of power, particularly focusing on how to deal with competing values and interests in processes of decision-making. Therefore, particular attention needs to be paid to power relations, whether material, economic, political, or cultural (Swyngedouw 2009). This would also include the analysis of the discourses and arguments that are mobilized to defend or legitimate particular strategies (Ibid.).

These questions of power come into play in an action situation where two or more actors are faced with a set of potential actions that jointly produce outcomes. Elinor Ostrom (2005, 32) refers to an action situation as the social space where participants with diverse preferences interact, exchange goods and services, solve problems, dominate one another, or fight (among the many things that individuals do in action arenas). The framework for power analysis, a particular application of the TSEI-framework presented in this book (Sect. 5.2), serves to shine a light on the political dimensions of Transformative Social-Ecological Innovation, using a vertical and horizontal typology of power.

The TSEI-framework has taken the action situation as the object of analysis and considers the action situation as the interface or 'glue' between two important analytical components: structure/institutions, on the one hand, and actor-agency, on the other (Huntjens et al. 2016). 'This relates directly to one of the important debates in social science: the relationship between structure and agency. Anthony Giddens (1984) argues that social structure is both the medium and outcome of action. According to Giddens (1984) and Alexander Wendt (1987), actors have preferences which they cannot realize without collective action; based on these preferences they shape and re-shape social structures, albeit also through unintended consequences and over a longer period of time (Grin 2010, 2011). Once these social structures are in place, they shape and re-shape the actors themselves and their preferences. In other words, the constitution of agents and structures are not two independent sets of phenomena, meaning that structures should not be treated as external to individuals. This is what Voß and Kemp (2015) call second-order reflexivity, which is about self-critical and self-conscious reflection on processes of modernity, particularly instrumental rationality. It evokes a sense of agency, intention, and change. Here actors reflect on and confront not only the self-induced problems of modernity, but also the approaches, structures, and systems that reproduce them (Stirling 2006; Grin et al. 2004). In other words, actors have the ability (agency) to evaluate the effectiveness of their actions in achieving their objectives. This means that if actors can reproduce structure through action, they can also transform it' (cf. Huntjens et al. 2016).

The discussion on structure–agency relationships has consequences for the interpretation of institutional change as put forward by many institutionalists. 'Although institutions may have a level of permanency, in our analysis of action situations the institutions are sustained or altered by the actions of the people that reproduce or change them. It is exactly at this juncture (i.e. in the action situation) that institutions are "renegotiated" and changed. When individual behaviour diverges from stated norms, structures will be renegotiated and may change. The duality of structure applies here: social structures determine and constrain social action, on the one hand, but are reproduced, renegotiated, or changed by that same human action simultaneously (Giddens 1984). Thus, institutional change is not a process by design, but by institutionalization' (cf. Huntjens et al. 2016). The process of institutionalization is referred to as follows: '[Institutions] are the outcome of a process of institutionalization, whereby preferred ways of doing things are progressively reinforced, making them relatively reliable. This process usually involves conflict and the exercise of social power' (Parker et al. 2003, 212). In this vein, Giddens' (1984) structuration theory, as well as the work of Bourdieu (1988, 2005) and Seo and Creed (2002), provides compelling arguments for depicting institutions not only as constraints on action, but also as the objects of constant maintenance or moderation. The example of TSEI-framework application in Sect. 6.6 shows when and how local agents change the institutional context itself, which provides relevant insights on institutional work (Beunen and Patterson 2019) and the mutually constitutive nature of structure and agency.

The duality of structure and complexities of institutional change are well explained by the concept of institutional contradictions (Seo and Creed 2002), a dialectic model emphasizing that multiple models of practice, conflicting structural rules, and contradictory principles among social agents are strong driving forces for organizational and institutional change. The model of Seo and Creed (2002) also emphasizes the role of less powerful or marginalized social actors as potential change agents. Finally, this dialectical model highlights 'the pivotal role of actors' ability or skills to mobilize institutional logics and resources from the heterogeneous institutional environments so as to legitimize and support their change efforts' (cf. Seo and Creed 2002, pp. 242). Hence, Seo and Creed (2002) argue 'that institutionally embedded praxis is a far more common and important factor in institutional change than institutional theories of either orthodox compliance or strategic resistance suggest'. Likewise, Beunen and Patterson (2019) have used the concept of 'institutional work' to explore the interplay between actors and institutional structures. The concept of institutional work is defined as the actions through which actors create, maintain, or disrupt institutional structures (Lawrence et al. 2009). However, Beunen and Patterson (2019) point out the difficulty of fully grasping an actor's real intentions and thus to distinguish purposive actions from other actions and communications that affect institutional structures, while recognizing that institutional structures are also influenced by a range of non-purposive actions taken by disparate actors. At the same time, actors are likely to have their own ideas about who played which role in the processes of institutional change. Hence, Beunen and Patterson (2019) argue for a broader definition of 'institutional work' by not only including the intentionality of actors, but also their non-purposive actions, and by recognizing that distinguishing purposive actions from other actions can be highly problematic. This require more attention for combinations of actions and strategies that can involve multiple kinds of linked actions. The TSEI analytical framework allows to zoom in on a series or cluster of related action situations (and their context), looking at 'structure' and 'agency' and at the output-outcomes-impact of these linked action situations (for more information see Sect. 5.1). The selected action situations are then analysed, focusing in particular on subcomponents such as initiation, process, format, and content of the action situation (Sect. 5.1).

At the core of TSEI lies the engagement and participation of government, businesses, academia, civilians, civil society, media, and the environment, in a process of multi-party deliberation, collective learning, and evidence-based decision-making, which resembles the quintuple helix innovation model (Barth 2011; Carayannis and Campbell 2010), which is a follow-up to the Triple Helix model, designed by Etzkowitz and Leydesdorff (2000). The Triple Helix focuses on the relations of universities, industry, and governments and is commonly used as a key concept guiding national and regional innovation policies around the world. This model for economic growth and regional development, however, has been criticized by many scholars, among others due to its lack of context sensitivity (see, e.g. Barth 2011; Carayannis and Campbell 2010; Williams and Woodson 2012). The quintuple helix innovation model (Barth 2011; Carayannis and Campbell 2010), in short 5-Helix, adds two important components: (1) the perspective of a media-based and culture-based public and (2) it frames knowledge and innovation in the context of the environment. The quintuple helix shows how democracy and the environment need to be integrated in the wider perspective of the architecture of Transformative Social-Ecological Innovation (TSEI) and societal transformation more in general.

In the Netherlands, for instance, the 5-Helix approach has become popular in the form of transdisciplinary approaches and living labs (see Sect. 4.12). Likewise, at Dutch universities of applied sciences, including my own university, it is becoming common practice to enhance cooperation within networks consisting of 5 components resembling the 5-Helix approach. In Dutch these components are called the 5 Os: Entrepreneurs (Ondernemers), Government (Overheid), Educational and Research institutions (Onderwijs- en Onderzoeksinstellingen), and Environment

(Omgeving), including citizens and civil society organizations. With these parties, it is possible to draw up core questions on relevant themes and set up programmes to answer these questions by means of applied research and educational programmes and projects. Together, the parties involved should commit themselves to finding the people and resources required. In this book, various methodological approaches of real-world experimentation and collaborative action research are being used that make the commitment to knowledge co-production operational (see Sects. 4.10 and 5.4 in particular).

For realizing a Natural Social Contract it requires a rethink of how society could be reorganized in such a way that more problems can be solved at the most appropriate level (the subsidiarity principle) and by new coalitions in horizontal innovation networks. It will require new forms of democracy, governance, organization, management, cooperation, changing laws and legislation, and a transition from linear to circular business models. It will go hand in hand with processes of collective learning (Sect. 4.8), in which different parties learn from each other and participate in joint knowledge development. Also innovative and hybrid forms of financing, such as revolving energy and sustainability funds, will be part of this development. A fundamental systemic change required for a Natural Social Contract, at least from an economic perspective, is a transition from mainstream economic thinking—with a singular focus on economic growth and financial profit—towards circular and regenerative economies and cultures (see Sects. 3.2 and 3.3), facilitated through shared value and multiple value creation (Sect. 4.9).

A Natural Social Contract may look similar to the Rhineland model in various ways, though a Natural Social Contract implies various systemic changes to improve the model's applicability to today's complex societal issues. At a more fundamental level, a Natural Social Contract provides a counter-proposal to the capitalist economic logic and divide between humans and nature, which is shaping the paradigms of both Anglo-Saxon and Rhineland models (see Sect. 3.5). Frequently mentioned drawbacks of the Rhineland model include its excessive emphasis on consultation, slow decision-making, lack for room for excellence ('mediocrity'), and ever-shifting goals due to an overestimation of the value of new insights (Bakker et al. 2005; Goodijk 2009; Peters and Weggeman 2009). A complementary approach to consultation between myriad interest groups, which is typical for the Rhineland model, is effective cooperation by new coalitions in horizontal innovation networks, also known as a coalition of the willing. Also hybrid forms of democracy, such as representative democracy complemented by deliberative democracy, could provide for better involvement of citizens and evidence-based decision-making on issues of common and public interest. New forms of governance, such as adaptive, reflexive, and deliberative governance, can help to increase stakeholder participation and commitment, community-involvement, policy learning, robust and evidence-based decision-making, flexibility and resilience to deal with shocks and surprises.

The TSEI-framework proposed in this book allows for a better understanding of and engagement with Transformative Social-Ecological Innovation (TSEI) in-the-making, not only focusing on how to govern the early stages of the process (e.g. transition arenas, niche-experiments), but also later phases of transition (for example, how to achieve acceleration, e.g. see (Gorissen et al. 2018; Sovacool 2016). This primarily includes research into the emergence, development and context of the partnership between various actors, and the extent to which they achieve the intended results. It requires research on what holds the actor coalitions together (e.g. shared beliefs and values, shared discourses, common interests, multiple value creation (Sect. 4.8) and procedural justice (Sect. 4.9), the roles of intermediaries in governing, facilitating, and accelerating transitions, and the role and influence of policy mixes (rather than studying single policy instruments). For more details on the research and innovation agenda see Part 3 of this book.

The **result** of the chosen procedure is de facto similar to the extent to which the overall goals of a Natural Social Contract are being achieved, with particular attention for TSEI outputs, outcomes, and impacts, and related to the ethical and normative aspects of a Natural Social Contract, such as social justice, equity, human security, environmental security, and planetary health. This requires development of appropriate indicators for measuring the social, ecological, and economic dimensions of sustainability transitions. Within this context, it is necessary to make a distinction between output, outcome, and impact (see Sect. 5.1 for more detail). The **output** could be a multi-party agreement and committed investments on Transformative Social-Ecological Innovation. Examples at the international level include the EU Green Deal, the UN Sustainable Development Goals, the Paris Climate Accord, or the Global Deal for Nature, yet to be realized at the biodiversity summit in Beijing in 2021. For other examples, at various levels, see Sect. 3.10. The **outcomes** are the direct effect(s) of the output. It is measurable and time-limited, though determining the full effect can take an extended period of time. Examples of outcomes include behavioural change, new knowledge, and (systemic) solutions resulting from co-creation and social learning. Specific examples constitute a circular (no waste) society and economy, preservation and restoration of biodiversity and ecosystems, and sustainable and healthy agri-food systems. Finally, **impacts** are the long-term or indirect effects of the outcomes and often difficult to quantify because they may or may not happen. Examples include achievement of the UN Sustainable Development Goals.

Transformative Social-Ecological Innovation (TSEI), as the engine of the sustainability transition, will help develop and implement a Natural Social Contract. Every single innovation that has the impact of a breakthrough or systemic innovation within the sustainability transition essentially constitutes a subsection of such a Natural Social Contract. Unlike hypothetical social contracts, however, TSEI can be studied empirically (see also Ziegler 2013). Chapter 6, therefore, provides various analytical instruments for empirical research into TSEI.

3.10 Development of a Natural Social Contract at Multiple Governance Levels

The development and implementation of a Natural Social Contract might take place at various governance levels, ranging from the local to the national and international level:

- **Local level**: At the local level it is often difficult to see systemic change in-the-making, although change is often initiated at the local level through niches or front-runners, for example, in pilot projects where local entrepreneurs, citizens, and/or other parties work together to put an innovative concept for sustainability into practice. When it comes to radically new practices, insights, and values, small steps can resonate, ultimately bringing about large-scale changes (Bryson 1988). In many cases, however, this requires strategic niche management (Kemp et al. 1998; Schot and Geels 2008), transition management or governance (see Sect. 4.2), or other types of long-term support and upscaling before systemic change can be consolidated. In this book several promising examples of niches and front-runners are provided in Chaps. 6 and 7, including Food Policy Councils (FPCs), as loci for practising food democracy, community-supported agriculture (CSA), as a sustainable alternative for industrial agriculture, and short food supply chains (Chap. 6), as well as circular business models, urban commons, and examples of eco-cities (Chap. 7).
- **Subnational level**: At this level, depending on the area or topic in question, there are generally more opportunities for systemic change, given that programmes and collaborations at this level usually require the involvement of multiple actors in a multi-level governance context. Examples include the development and implementation of new forms of spatial and participatory planning processes for sustainable cities or river basin management or a transition approach towards sustainable agriculture at the provincial level. Illustrations in this book include 'The most sustainable square kilometre of the Netherlands' (Sect. 7.1), and the transition approach towards sustainable agriculture in the Province of South Holland (Sect. 6.6). Also the regional cooperative 'Land of Values' (In Dutch: *Land van Waarde*) serves as an example at the subnational or supra-local level (Sect. 6.3).
- **National level**: A societal transformation towards a Natural Social Contract at the national level is a complex multi-level governance challenge, requiring the fine-tuning of top down policy and visions with important bottom-up processes (Huntjens et al. 2011a, b; Bache et al. 2016). Sustainability transitions at the national level are certainly not only initiated or facilitated by government, but also driven by the private sector, civil society organizations, knowledge institutes, and public opinion, as explained in the previous section. Sustainable development, in particular the Sustainable Development Goals (SDGs) and the Paris Climate Agreement, has entered public opinion, domestic laws and policies in different ways and to different degrees in almost every country, even though both agreements are not legally binding but emphasize consensus-building and

voluntary and nationally determined targets. As a consequence, the 195 countries that signed the Paris Climate Agreement have not yet devoted enough effort to reducing their greenhouse gas emissions to limit global warming to a maximum of 1.5 degrees Celsius (IPCC 2018). The UN climate panel notes that global warming is currently more likely to reach 3 degrees than 2, let alone the targeted 1.5 degrees. At the same time, domestic laws and policies are influenced due to a country's signature to international environmental agreements or treaties that are legally binding, such as the UN Convention for the Law of the Sea (UNCLOS) on the conservation and sustainable use of marine biological diversity, the adoption of the Kigali amendment to the Montreal Protocol (to reduce emission of hydrofluorocarbons (HFCs)), international law of freshwater, or the integration of environmental considerations into investment, trade and intellectual property law (Dupuy and Viñuales 2018). In parallel, domestic policies and regulations are evolving and adapting in response to public opinion, advocacy, court rulings or parliamentary decisions, of which the latter is heavily influenced by political constellation and democratic processes. As such, a growing number of countries have recognized Rights of Nature in their legal frameworks and/or jurisprudence. Legal provisions recognizing the Rights of Nature, sometimes referred to as Earth Jurisprudence, include constitutions, national statutes, and local laws, e.g. in Uganda, Peru, Ecuador, Colombia, India, Bangladesh, New Zealand, and communities in the USA, while some countries have granted rights to rivers, e.g. the Whanganui river in New Zealand, the Yarra River in Australia, and the Ganges and Yamuna rivers in India. Constitutional amendments addressing, among others, the rights of the living, animal welfare, the global commons, the crime of ecocide, and the principle of non-environmental regression have been tabled in many parliaments, signalling a trend for a more Earth-centred constitutional process.

- **EU level**: In Europe, 'a good and healthy life in 2050 within our planet's ecological boundary' is a core component of environmental policy (EU, 7th Environment Action Programme 2013). This vision has also been incorporated in other lines of EU policy. In the past two decades, the European Union has introduced a large body of environmental legislation, which has succeeded in significantly reducing air, water, and soil pollution. More recently, in 2019 the new European Commission has announced a New Green Deal for Europe, which is an ambitious and pragmatic plan to transition to zero greenhouse gas emissions and transform Europe in the process. The proposed 'Green Deal' represents a unique opportunity for the EU to move away from fragmented policymaking in climate change to a comprehensive and consistent policy framework. This can promote decarbonization while also taking advantage of the economic and industrial opportunities it offers, such as circular economy, clean energy, and related job creation. In a climate-neutral Europe, all industries relying on burning fossil fuels will have to change to cleaner and renewable energy sources. At the heart of the Green Deal the Biodiversity and Farm to Fork strategies point to a new and better balance of nature, food systems, and biodiversity.

From a legal perspective, 'modern European Union (EU) legal frameworks treat Nature as property and implicitly legalize damage through regulations which treat ecosystems as objects and not subjects of law. Traditional environmental regulatory systems generally describe nature as property to be used for human benefit, rather than a rights-bearing partner with which humanity has co-evolved. Civil society organizations have proposed changes in EU legal frameworks to account for Nature's Rights, since the purpose of the existing regulations is to establish how much damage can be done and not to prevent it and/or eradicate it' (cf. Pikramenou 2020).

- **Global level**: 'International processes associated with sustainable development have not led to an internationally legally binding framework that adequately addresses the challenges we face' (cf. Rühs and Jones 2016). Nevertheless, important developments on a global scale include the adoption of the United Nations Sustainable Development Goals (SDGs) and the Paris Agreement on Climate Change, although none of them is legally binding. Both agreements emphasize consensus-building and allow for voluntary and nationally determined targets. Likewise, a global nature agreement is currently being worked on, which should lead to a Global Deal for Nature at the biodiversity summit in Beijing in 2021. This is an important step towards a major reorganization of the entire economic and financial system, a global shift towards sustainability that must go hand in hand with the fight for the preservation of biodiversity and the battle against climate change. In general, values on the relationships between humans and nature are becoming more prominent and recognized at the international level in various ways. For example, the United Nations World Charter for Nature, adopted in 1982, announced five principles of conservation by which all human conduct affecting nature is to be guided and judged. Likewise, the Earth Charter—an ethical framework for sustainable development published in 2000—reserves a central place for environmental protection, human rights, equitable human development, and peace and argues that these values are interdependent and indivisible. In 2010, the Universal Declaration of the Rights of Mother Earth was proclaimed at the World People's Conference on Climate Change and the Rights of Mother Earth held in Cochabamba, Bolivia. Overall, evidence suggests a growing presence of international environmental law in international legal practice (Dupuy and Viñuales 2018).

The sustainability transition implies a large-scale societal transformation towards a Natural Social Contract, in which Transformative Social-Ecological Innovation (TSEI) will be needed in different fields and at different levels of scale. These systemic innovations may occur both simultaneously and independently of each other, reinforcing each other or competing with each other. As a rule, such a transition is not a linear development but consists of a mosaic of various technological and social innovations, of which we cannot predict in advance what will and what will not work.

Open Access This chapter is licensed under the terms of the Creative Commons Attribution 4.0 International License (http://creativecommons.org/licenses/by/4.0/), which permits use, sharing, adaptation, distribution and reproduction in any medium or format, as long as you give appropriate credit to the original author(s) and the source, provide a link to the Creative Commons license and indicate if changes were made.

The images or other third party material in this chapter are included in the chapter's Creative Commons license, unless indicated otherwise in a credit line to the material. If material is not included in the chapter's Creative Commons license and your intended use is not permitted by statutory regulation or exceeds the permitted use, you will need to obtain permission directly from the copyright holder.

Part II

Theories and Concepts

Conceptual Background of Transformative Social-Ecological Innovation

In this chapter I survey key theories and concepts that provide substance to the workings of Transformative Social-Ecological Innovation (TSEI). A number of relevant theories and concept have already been mentioned in the previous chapters, such as Social Contract theory (Sect. 3.1), and in Sects. 3.8 and 3.9, including resilience theory and social-ecological systems (Sect. 3.8), quintuple helix innovation model (Sect. 3.9), as well as institutional change and the structure-agency debate (Sect. 3.9), and several economic theories (Sects. 3.2 and 3.3). In this chapter I will start with providing a conceptual discussion and definition on Transformative Social-Ecological Innovation (Sect. 4.1), and devote more attention to various theories and approaches that are relevant for TSEI, such as transition studies (Sect. 4.2), institutional design principles for governing the commons (Sect. 4.3), design principles from nature (Sect. 4.4), complex adaptive systems (Sect. 4.5), adaptive, reflexive, and deliberative approaches to governance, management, and planning (Sect. 4.6), social learning, policy learning, and transformational learning (Sect. 4.7), shared value, multiple value creation, and mutual gains approach (Sect. 4.8), effective cooperation (Sect. 4.9), transdisciplinary cooperation, living labs, and citizen science (Sect. 4.10), and the art of co-creation: approaches, principles and pitfalls (Sect. 4.11).

Drawing on the insights from this literature, I argue that studying Transformative Social-Ecological Innovation should involve a look at both structure and agency, in particular at decisive moments where both structure and agency intersect (i.e. in action situations), as well as the resulting outputs, outcomes, and impacts. I identify a critical need for attention to the fundamentally political character of Transformative Social-Ecological Innovation, and the need for multiple value creation, in which parties seek for shared values, mutual gains, and common interest.

The original version of this chapter was revised. Incorrect citation of Fig 4.5 in page 100 has been corrected to read as Fig. 4.3. The correction to this chapter is available at https://doi.org/10.1007/978-3-030-67130-3_9

4.1 Definition of Transformative Social-Ecological Innovation (TSEI)

In this section I introduce the concept of Transformative Social-Ecological Innovation (TSEI), as a response to the limitations of more traditional and anthropocentric notions of social innovation, in particular when applied to sustainability issues.

Social innovation is not a new idea but has become immensely more popular in the past two decades, not least because traditional models of innovation—targeting technological innovation—often fall short in a world of complex societal challenges. A technological innovation with social impact is, therefore, not the same as a social innovation. Social innovation mainly consists of new forms of governance, organization, management, participation, and cooperation within and between government, businesses, citizens, civil society organizations, and research and education institutions.

Within the Social Sciences, there are various definitions of social innovation. Haxeltine et al. (2016:20) define a social innovation (SI) as a change in social relations, ushering in new ways of thinking, doing, and organizing. If this social innovation can also bring about a systemic change in a specific context, it is called a Transformative Social Innovation (TSI). Transformative Social Innovation (TSI) is conceptualized as social innovation that challenges, alters, or replaces dominant institutions in the social context (Haxeltine et al. 2016). Another definition is provided by Moulaert et al. (2013), in which social innovation is transdisciplinary and defined as a social, innovative process, along with its outcome, that contributes to:

- fulfilling human needs;
- creating and strengthening social relationships;
- enhancing the socio-political capacity of citizens.

Several points of criticism can be levelled at this definition, however, such as a one-sided focus on human needs. In contrast, for addressing sustainability issues, the intricated coupling between human and biophysical systems needs to be recognized. In other words, it is not only about human needs but also about the needs of nature, life-supporting ecosystems, and our planet, on which humans depend. Enhancing the socio-political capacity of citizens, in Moulaert's definition, is also one-sided, as social innovation is not only about 'citizens' capacity', but about enhancing 'society's capacity' to innovate, which involves effective cooperation between multiple parties within society (see quintuple helix innovation model in Sect. 3.9).

Haxeltine et al. (2016) moreover criticize a definition of social innovation that also describes the outcome, because this can lead to processes in which the destination transcends the journey towards it. To avoid this, they define social innovation as a process rather than as a result.

The Dutch Advisory Council for Science and Technology Policy (2014) defines social innovation as: 'New solutions (products, services, models, markets, processes, etc.) that simultaneously meet a societal need (more effectively than existing solutions) and introduce or improve capacities and relationships and a better use of resources'. In other words, social innovations are good for society and increase its capacity for action.

When applying the concept of social innovation to sustainability issues, it becomes clear that the concept, by definition, is limited by its anthropocentric approach, instead of recognizing the intricate coupling of human and biophysical systems. In none of the above anthropocentric definitions there is a mention of ecosystems, neither an acknowledgement of the relation between social, economic, and ecological systems. At best, (natural) resources are considered to be used exclusively by humans, to serve the needs of humanity, as in the definition by the Dutch Advisory Council (2014), which reserves a central place for the 'societal need'. Olsson and Galaz (2012), therefore, argue that addressing only the social dimension will not be sufficient to guide society towards sustainable outcomes. Within this debate, the concept of eco-innovation was introduced by René Kemp and Peter Pearson (2007) as a valuable contribution to better understand innovation targeted at solving sustainability challenges. In the literature, the term eco-innovation is generally understood to mean 'the production, application, or exploitation of a good, service, production process, organizational structure, or management or business method that is novel to the firm or user and which results, throughout its life cycle, in a reduction of environmental risk, pollution, and the negative impacts of resource use (including energy use) compared to relevant alternatives' (Kemp and Pearson 2007).

As a response to above anthropocentric definitions, and in addition to the concept of eco-innovation by Kemp and Pearson (2007), Per Olsson and Victor Galaz (2012) introduced the concept of Social-Ecological Innovation, which is defined as 'social innovation, including new technology, strategies, concepts, ideas, institutions, and organizations that enhance the capacity of ecosystems to generate services and help steer away from multiple earth-system thresholds'. Although this definition by Olsson and Galaz does acknowledge the intricate coupling between social and ecological systems, it does not explicitly mention the fundamental and systemic changes that are required in social and economic systems, such as a transition to regenerative and circular economies and cultures, in order to realize a sustainable and regenerative society. Hence, this definition could be improved by adding an explicit recognition for the need of systemic innovation, defined as 'profound transformations in social systems', which involve 'changes in established patterns of action as well as in structure, which includes dominant cultural assumptions and discourses, legislation, physical infrastructure, the rules prevailing in economic chains, knowledge infrastructure, and so on' (Grin et al. 2010).

Following from above, in this book I propose to complement and redefine the concept of Social-Ecological Innovation (SEI) by Olsson and Galaz (2012), with the concept of Transformative Social Innovation (TSI) by Haxeltine et al. (2016), resulting in the concept of 'Transformative Social-Ecological Innovation' (TSEI). This concept is applicable to systemic innovations in social-ecological systems related to water, food, energy, biodiversity, climate change, health, spatial planning, mobility, and built environment, as well as a transition to regenerative and circular economies and cultures. The overall goal of TSEI would be to realize a Natural Social Contract that promotes human security, social justice, and planetary health (see Sect. 3.8).

Following the above reasoning, I define Transformative Social-Ecological Innovation as 'systemic changes in established patterns of action as well as in structure, including formal and informal institutions and economies that contribute to sustainability, health, and justice in all social-ecological systems' (definition by author). It is about society re-asserting a sustainable, regenerative, and healthy future. To avoid confusion about the term 'institution' I use the definition proposed by Calhoun (2002, p.33): 'Institutions are deeply rooted patterns of social practices or norms that play an important role in how society is organized'. Institutions can pertain to various areas of social activity, such as family life, associations, and politics. Generally speaking, institutions result from a process of institutionalization, in which preferences are gradually strengthened until they are fixed and familiar. This process is usually accompanied by conflicts and the exercise of social power (Parker et al. 2003). Following the definition of Calhoun (2002) I distinguish between formal and informal institutions, where the first includes the constitution, laws, and legislation, and the latter includes customary law, existing practices, norms, and culture (see also Sect. 5.2).

The definition of Transformative Social-Ecological Innovation (TSEI) proposed here is in line with the critical tradition of social innovation within sociology, which argues that such innovations aid in a transition away from the current regime rather than lead to reforms within it. Contrary to this sociological interpretation, an increasingly economic interpretation of social innovations has gained ground in recent years, which primarily defines social innovation as an instrumental, helpful tool to ensure the continuation of common developments by improving what already exists (Dagevos 2018). In this sense, social innovation is absorbed into the prevailing paradigm, ridding social innovation of its rebellious, socially critical character, as well as its transformative power (*ibid.* 2018). For paradigm shifts such as the sustainability transition, it is important that Transformative Social-Ecological Innovations remain 'radical'. This, however, can be very difficult to accept for a great many parties, i.e. the powers that be. Consequently, issues of distribution and power play an important role in societal transformation processes (Meadowcroft 2009; Cattacin and Zimmer 2016; Karré 2018; Huntjens 2019), because it involves several groups, each of which have their own norms, values, and interests. This also brings mechanisms of inclusion and exclusion into play (Sassen 2014), which can lead to conflicts and lawsuits. It will always be a battle to overcome vested interests, change existing systems and paradigms. In Sects. 4.1 and 5.3, I will follow-up on the critical need for attention to the fundamentally political character of Transformative Social-Ecological Innovation, and the need for multiple value creation, in which parties seek for shared values, mutual gains, and collective well-being in a social-ecological setting.

4.2 Transition Studies

For more than 20 years, transition studies have been an interdisciplinary research field focusing on the complexity of changes in systems in society. It is interdisciplinary because the field takes a holistic approach to analysing society, including social, cultural, institutional, technological, ecological, economic, and political aspects. Transition theory is particularly useful in identifying solutions to wicked policy problems (Avelino et al. 2016, p. 557; Rittel and Webber 1973). These problems are so persistent that inter-party cooperation is essential if a solution is to be found. Clashes of values are inevitable, as the sustainability transition calls for fundamental changes to be made in different layers of society. At the same time, the power to influence this process will often be divided among several actors who may perceive problems differently (Grin et al. 2010; Kemp et al. 2007, p. 316).

The field of transition studies has brought forth a multitude of theoretical frameworks, such as the Multi-Level Perspective (MLP) model, a theoretical framework for analysing systemic transitions over an extended period of time (20–40 years) (Geels 2011; Geels and Kemp 2000; Smink 2015). See Fig. 4.1.

The MLP states that transition arises as a consequence of interactions between three analytical levels. At the micro-level, innovative practices, or niche experiments, emerge. These are innovative social, economic, technological, or policy practices that are different from and often protected from the dominant regime. The concept of a niche can be equated to that of social innovation when both provide a collective, new solution to a societal problem or social need. In practice, the meso-level is most often the structural context of such developments, encompassing the dominant culture, formal and informal rules, routines, knowledge, and infrastructure that perpetuate a particular practice.

At the meso-level, for instance, it is interesting to explore TSIs implemented by the regime/powers that be. The macro-level is the landscape on which major changes take place in terms of politics, culture, and worldviews, e.g. globalization and individualization, or natural features that can be difficult to influence and tend to change slowly. Landscape developments are the result of the ideas and actions of large numbers of players, as it takes very long for actors to influence the landscape because of its tendency for slow change. Examples include demographic and economic trends and political ideologies (Geels and Schot 2007; Loorbach and Rotmans 2006), as well as ecological processes and major disasters such as Fukushima, and the impact they have on the system.

The chances of and possibilities for systemic changes increase as the regime starts to experience more pressure from the landscape and niches, which may ultimately open up a window of opportunity. Once a niche is sufficiently developed it can break through in the regime. When a niche or social innovation can bring about a systemic change, it is often called a Transformative Social Innovation (TSI), systemic innovation or transition. However, transitions are rare because they require exactly the right interaction between the landscape, regime, and niche (Grin et al. 2010, p. 328).

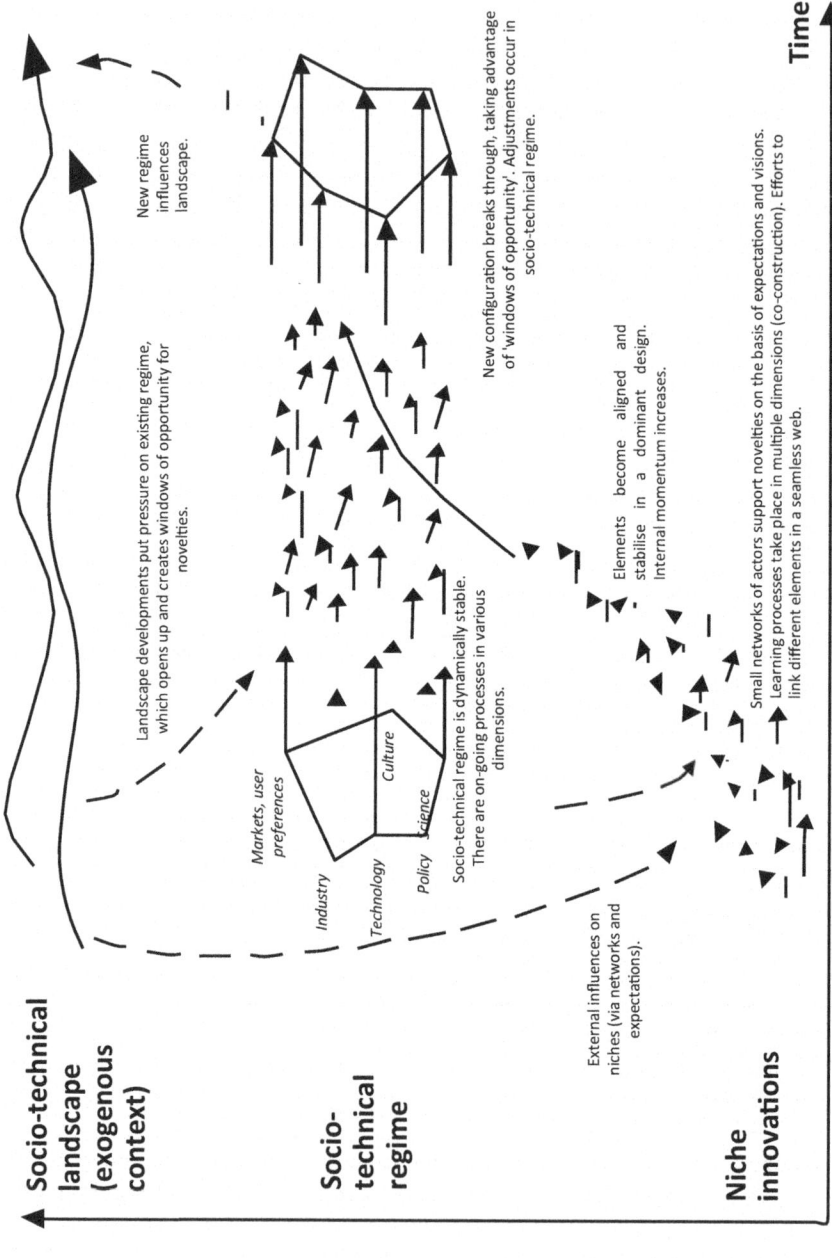

Fig. 4.1 The Multi-Level Perspective according to Geels and Schot (2007)

According to Grin, stating that only landscape pressure has the capacity to change policy is overly simplistic. After all, the very politics behind the emergency of new policy and the regime change are crucial if we are to properly understand transitions (Grin 2011, p. 7). Moreover, attempts to change established patterns always come up against resistance, rigidity, and/or normative questions as to the legitimacy, justness, methods, and direction of the transition (Grin 2016, p. 112; Meadowcroft 2009).

Within transition studies, we can distinguish several different governance approaches and philosophies, including:

1. Strategic niche management (Kemp et al. 1998).
2. Transition management (Rotmans et al. 2001; Loorbach 2010).
3. Reflexive governance (Voß and Bornemann 2011).

Whereas strategic niche management (SNM) focuses more on developing new innovations, transition management (TM) is more geared towards strategic interventions implemented by and within the established order. TM is a governance paradigm that sees complexity and uncertainty as drivers of social innovation rather than regarding them as difficult obstacles that must be controlled or managed. Since 2000, TM has been a widely used policy strategy in Dutch ministries. The governance perspective places greater emphasis on policy and the political process within transitions than the other versions do (Grin et al. 2010; Loeber 2003; Rotmans et al. 2001), thereby providing relevant insights for research on Transformative Social-Ecological Innovation (TSEI) conducted by this research group. SNM and TM are both concrete governance philosophies, with TM going even further than SNM. Both philosophies go very far in creating perspective for action, though they are sometimes criticized for their somewhat 'modernist' character. Reflexive governance and reflexive monitoring are much more modest in this respect, are less based on control, and are not as clearly designed with an eye on MLP.

Several academics have criticized MPL and TM for ignoring the politics behind transitions (Avelino et al. 2016; Kern and Alber 2006; Meadowcroft 2009). In recent years, however, multiple articles have been published that have included power in their description of the dynamic transition process (Avelino et al. 2016; Avelino and Rotmans 2009; Hendriks and Grin 2007; Kern and Smith 2008; Meadowcroft 2007, 2009; Shove and Walker 2007). Meadowcroft (2009, p. 329) believes it is risky for policymakers to underestimate the political dimensions of transitions. It must be noted, of course, that this is often a difficult task for policy officers because they are executive officers, while elected officials usually determine the (political)goals. There is always a normative aspect to the goal of a transition (see Sect. 3.8). What direction should society take? A good example is the modernization of agriculture in the twentieth century, which many people believed to be a worthwhile objective. However, the agricultural industrialization would later go on to cause problems due to monocultures and the use of pesticides (Hendriks and Grin 2007). Ultimately, the direction of this 'agricultural modernization' was not 'neutral' or 'technically optimal' as much as it was a political choice. In much the same way, the transition to climate-proof cities will raise normative questions on what makes cities climate-proof and who should bear the costs involved in the process (Eriksen et al. 2015).

According to Meadowcroft, social and political conflict about the nature and direction of a transition are inevitable, if only because priorities will have to be set and government budgets are limited, although the government can also use legislation to steer developments in the right direction. Moreover, everyday cultural and political phenomena also play a role in determining which policy choices are made (Meadowcroft 2009, p. 326). When normative conflicts and conflicting interests provoke discussion, the aforementioned political dimensions of transitions can be seen clearly.

Grin has taken Meadowcroft's criticism to heart, adapting his approach to the practical aspects of policymaking with concepts such as reflexive governance and dual-track governance (Avelino et al. 2016; Grin 2016; Hendriks and Grin 2007). These methods were designed to help policymakers take the political dimensions of transitions into account and to respond to them strategically. Reflexive governance helps actors to critically reflect on (the process of) their desired transition. Dual-track governance discusses how actors can make strategic connections between the different levels (Hendriks and Grin 2007).

4.3 Institutional Design Principles for Governing the Commons

Commons are natural or cultural resources that are available to all members of a group or society, such as shared fishing waters, forests, and agricultural land, as well as sources of information, knowledge, and culture. Increasingly, citizens and local organizations are opting for joint management of commons instead of private ownership. In many instances, the joint management of commons is considered as a kind of correction mechanism for the economic climate of liberalization, privatization, and individualism in recent decades. However, it is quite unlikely that in advanced societies, resource users will be able to govern their exploitative action all by themselves, without the help of other socio-political agents and agencies, including the state.

The problem of the commons today is that we still tend to think of it as a common resource, whether it be oceans and rivers or fish stocks and grazing lands or neighbourhoods and cities or the Internet and social media. This is a misunderstanding. Because the joint and sustainable management of commons cannot succeed without institutions for collective action. Elinor Ostrom argued that the commons require a set of rules. She won the Nobel prize in economics for proving that these resources need not succumb to the so-called tragedy of the commons (exploitation by someone taking more than their share) if a system of checks and balances prevails.

In 1990, Elinor Ostrom published eight institutional design principles for the joint, sustainable management of commons. Ostrom defines commons as a social practice of governing a resource not by state or market but by a community of users

4.3 Institutional Design Principles for Governing the Commons

that self-governs the resource through institutions that it creates. Ostrom spent her entire life collecting evidence about the management of commons all around the world, and her research provides convincing evidence that people can succeed to sustainably manage public goods. Ostrom has demonstrated that self-organizing communities can, indeed, manage common pool resources in a sustainable fashion and does not necessarily need to be regulated via a central governing mechanism. Ostrom's work generated an approach that can be used in the analysis and design of effective institutions (or instruments) to manage not just common pool resources but many different types of shared resources (Foster and Iaione 2018).

According to Ostrom, there are 8 institutional design principles for the sustainable management of shared resources:

- Clearly defined boundaries (what are the commons and who own them).
- Adaptation to local conditions.
- Joint decision-making by owners.
- Supervision by or on behalf of the owners.
- Penalties for misuse.
- Low-cost and easily accessible arbitration in the event of disputes.
- Community self-management and recognition by higher authorities.
- For large-scale commons, a layered system with local groups.

Many studies have since explicitly or implicitly evaluated these design principles. Ostrom's first principle, for instance, pertains to the demarcation of clear boundaries around a community of users and the system of resources that this community uses (Agrawal 2002), and has been the principle to attract the most, mainly theoretical, criticism (Cox et al. 2010). The primary complaints about this principle target its excessive rigidity, stating that social or geographical boundaries cannot or need not be defined very clearly in most systems, mainly to allow for more flexible, ad hoc arrangements between participants (ibid.).

Huntjens et al. (2012) show that dealing with complex societal challenges, such as climate change adaptation, requires a number of adapted and additional design principles, including a robust and flexible process, adaptive planning and mechanisms for social learning and policy learning. Table 4.1 presents a brief overview of these adapted and additional design principles (ibid.), which correspond to elements of sustainability learning (see, for example, Beers et al. 2016). Further research is needed in order to test principles on their usefulness and to make them more applicable to TSEI. Empirical research by Termeer et al. (2013) and Runhaar et al. (2017) shows the potential and pitfalls of self-organization and the sustainable management of agricultural and natural land by agricultural cooperatives in the Netherlands, emphasizing the complicated and ambivalent nature of interactions between farmer cooperatives, nature organizations, and government.

Emphatically, these design principles are not a blueprint for action, but have been formulated in such a way that they can be adapted to local and regional contexts of specific geographies, ecologies, economies, and cultures. For example, is it possible to apply these institutional design principles to cities to rethink the governance of

Table 4.1 Institutional design principles relevant for TSEI aimed at sustainable management of shared resources, such as community-supported agriculture (based on Huntjens et al. 2012 and Ostrom 2005)

Institutional design principle	Explanation
Adaptive, reflexive, and deliberative approaches to governance	Governance taking account of ambivalence, complexity, uncertainty, and distributed power in societal change.
Equal and fair (re-)distribution of risks, costs, and benefits	Through the involvement and strong representation of groups and stakeholders who will be affected or are particularly vulnerable.
Arrangements for collective decision-making	To enhance the participation of groups and stakeholders in decision-making processes.
Reflexive monitoring	This provides a foundation for reflection and social learning, while at the same time supporting accountability.
Conflict prevention and resolution mechanisms	Prevention and resolution of conflicts is possible through a variety of mechanisms, such as appropriate benefit sharing arrangements, mutual gains approach (see Sect. 4.8), timing and careful sequencing, transparency, building trust, and sharing or clarifying tasks, powers, and responsibilities.
Embedded activities/polycentric governance	Governance and management at a level of scale that does the most justice to the complexity of socio-ecological systems. For example, in European law this is similar to the principle of subsidiarity: social and political issues should be addressed at the most immediate or local level.
Policy learning	By exploring uncertainties, considering alternatives and 'reframing' problems and solutions, as well as policy experimentation: a deliberate and coordinated activity (e.g. pilot projects) to develop and test new policy alternatives.

cities and the management of their resources? The answer is negative, since these principles cannot be simply copied to the city context without significant modification. Hence, it is important to stress that design principles such as those presented here are never interpreted or used as a panacea, as they are primarily intended to create more insight into and awareness of various aspects of collective action, though they can be used in practice as entry points for the organization and support of social innovation, collective action, and multi-party collaboration.

For the sustainable management of shared resources the institutional design principles for governing the commons are increasingly being used in various ongoing sustainability transitions, as highlighted in Chaps. 6 and 7. Examples within the food transition include food networks, citizens' farms, community-based agriculture, and short food chain initiatives, in which farmers and consumers work together to produce their own food without relying on wholesalers and supermarkets (see Sects. 6.3 and 6.4). Similar developments can be observed in the urban context, where the concept of 'urban commons' is gaining popularity (Colding et al. 2013; Bollier and Helfrich 2015; Foster and Iaione 2018). 'This constitutes a growing number of urban commons showing that it is not only possible but highly attractive to create commons through which citizens can actively participate in the design of their city spaces and the programmes and policies that govern them' (cf. Bollier and Helfrich 2015). However, Foster and Iaione (2018) argue that Ostrom's principles cannot be simply adapted to the city context without significant modification. First of all, natural commons (e.g. fish stock, forests, or grazing lands) are different from urban commons (e.g. community gardens, parks, neighbourhoods, urban infrastructure, or the whole city as a commons). After analysing 200 urban commons in 100 cities, Foster and Iaione (2018) propose 'a set of design principles that are distinctively different from those offered by Elinor Ostrom and which can be applied to govern different kinds of urban commons, and cities as commons'.

From a commons perspective, a number of fundamental and systemic questions can be postulated to ongoing transitions, such as the transition to a sustainable and healthy agri-food system (Chap. 6) and the urban sustainability transition (Chap. 7). Some examples of such questions include:

- What if agricultural land or city spaces would be managed as a public commons, with user rights instead of property rights? Could this avoid land speculation and selling land to the highest bidder only? Could this avoid excessive mortgages and financial risks for farmers and food pioneers? And would it result in a more sustainable management of shared resources, with fair prices for the services provided by those commoners? For instance, how can we pay a fair price for food products and producers in a society that expects sustainable food production, which at the same time contributes to public health, animal welfare, climate change adaptation and mitigation, biodiversity, nature conservation, etc.?
- How does a group or a community, who's members are in an interdependent situation, organize and govern themselves to obtain continuing joint benefits from the collective management of commons? In this book the example of the regional cooperative 'Land of Values' (In Dutch: LandvanWaarde) is highlighted in Sect. 6.3. And what is the role of government given this self-organizing capacity of commoners? 'The task of governments in contemporary, complex societies is to influence social interactions in such a way that political governing and social self-organization are made complementary' (cf. Kooiman 1993, p. 256), and from this perspective it is perhaps more realistic and effective to co-govern, in particular within a heavily regulated policy field such as agriculture (e.g. see Termeer et al. 2013). From a governance point of view, the challenge is to formulate a role for the government, which encourages rather than discourages self-organizing activity.

4.4 Design Principles from Nature: Benchmarks for a Natural Social Contract

Seeing a Natural Social Contract as the object of this book, it is only logical to pay extra attention to design lessons from nature. From an analytical perspective, these natural design principles can be used as a benchmark for a transformation towards a sustainable, healthy, and just society. The insights presented in this chapter, and particularly those pertaining to complex adaptive systems (Sect. 4.5) and adaptive governance, management and planning (Sect. 4.6), as well as the institutional design principles for sustainable management of commons (Sect. 4.3) are, in many respects, similar to design lessons taught by nature, such as adaptive capacity, resource efficiency, circularity, and self-organization. This, however, is hardly surprising when we consider that most of the literature reviewed here came from the field of ecosystem management (Holling 1978; Walters 1986; Pahl-Wostl 1995; Lee 1999).

In 1997, Janine M. Benyus published the book 'Biomimicry: Innovation Inspired by Nature', in which she introduced the concept of biomimicry, translating lessons from nature into our daily lives, to a wider audience. In the two decades since her book was published, the concept of biomimicry has gained in popularity, as has, more recently, the interest in biomimicry in the context of social innovation, with the aim of creating products, processes, and policies that are well-adapted to life on earth in the long term. Benyus defines biomimicry as 'a new science that studies nature's models and then imitates or takes inspiration from these designs and processes to solve human problems, e.g. a solar cell inspired by a leaf' (Benyus 1997).

While recognizing the revolutionary character of biomimicry as a design concept for human systems of production, Mathews (2011) argues for a deeper philosophy of biomimicry. It is not only about our production systems that need to be adapted, but also our consumption patterns, labelled by Mathews (2011) as 'psychocultural patterns of desire'. Mathews argues that biomimicry will remain limited as a pathway to sustainability when acting only in imitation of nature, but requires acting from within the mindset of nature.

Within a Natural Social Contract, biomimicry should not be taken as an 'imitation' of life as much as a 'return' to natural, sustainable behaviour by humankind as a component of a greater ecosystem, that of planet earth. However sad this observation may be, the concept of biomimicry alone confirms that humankind lost its way at some point and stopped seeing itself as something 'natural', to the point that we are now forced to mimic nature in order to survive as a species and to ensure the planet remains liveable for future generations. That is why I would argue that a Natural Social Contract does not constitute mimicking nature, but rather constitutes a return to our origins.

Biomimicry is based on Life's Principles (see Fig. 4.2), which are certain design lessons from nature based on general patterns and strategies found among the myriad species that live and flourish on earth. Similar to biomimicry, a Natural Social Contract assumes that all life on Earth is interconnected and interdependent, as well as dealing with the same set of conditions (sunlight, water, gravity, cyclical processes, complex systems, non-linear feedback loops, etc.). Life on Earth has developed a series of strategies over the past 3.8 billion years that are optimized for these conditions and to enable life. For instance, there are innumerable strategies in nature to use scarce resources in a smart way. By learning from these natural design lessons, we can develop innovative strategies and test our institutional and economic system designs against these sustainability benchmarks.

4.4 Design Principles from Nature: Benchmarks for a Natural Social Contract

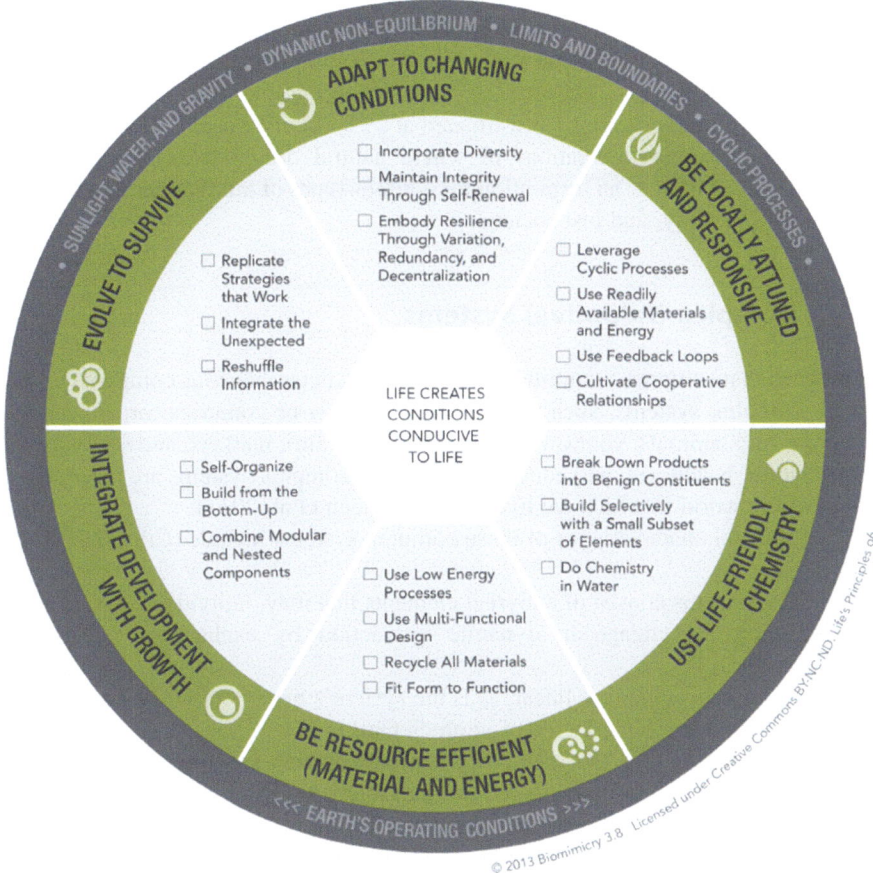

Fig. 4.2 Design principles from nature (Biomimicry 3.8, 2013)

For instance, there are innumerable strategies in nature to use scarce resources in a smart way. Circular Economy and Regenerative Economy are examples of economic design based on ecology, where nature shows how circularity is usually organized at the lowest possible level. For companies it will require a fundamental shift from linear to circular business models. In this respect, a company could be compared with an organism in an ecosystem. When the system changes, it is smart to take a close look at energy, water, and material flows. This offers opportunities for a company to save money, innovate, and deploy new strategies. Or the company could even fulfil a whole new function within the economy. In nature, there are many animal species that break down organic material such as dead leaves into humus. The organic material is broken down by the bugs so far that the trees can take up the raw materials from the humus again and reuse them to make new leaves. Recycle companies in a circular economy have a similar function. However, there are not that many types of it. A large part of our clothing, furniture, electronics, and our food ends up at a dumping ground for waste or is incinerated, while it still contains a lot of valuable raw materials. Much more can be done, but getting there requires technological and social innovation, including organizational and institutional innovation. The ratio between the impact of technology and social innovation for realizing a circular economy is estimated at 25:75 (Jonker et al. 2018). Table 4.2 below provides some illustrations of where natural design principles could be translated into, or used as an inspiration for, various types of innovations that support a sustainable, healthy, and just society.

4.5 Complex (Adaptive) Systems

All systems that comprise a multitude of interlinked, heterogeneous components are, in fact, complex systems, such as ecosystems, cities or countries, organizations, organisms (e.g. animals, your own body, and your brain), markets, and sectors such as agriculture, healthcare, and education. As such, complex systems are a pervasive feature of the world in which we live (Van der Steen et al. 2011).

Some general characteristics of these complex systems include (Cilliers 2000):

- Complex systems consist of a myriad elements that may, individually, be simple.
- These elements engage in dynamic interaction by exchanging energy on information.
- These interactions are non-linear, as is the system's behaviour as a whole, which means they have a high degree of unpredictability.
- Complex systems are emergent: new properties, patterns, regularities, and/or completely new entities are created through interaction.
- There are many direct and indirect feedback loops.
- Complex systems are open systems: they exchange energy or information with their environment.
- Complex systems have memory, not in a specific place, but distributed throughout the system. Each complex system, therefore, has a history and an evolutionary character, which helps determine how the system behaves.
- The system's behaviour is determined by the nature of the interactions within it, not by what is in its components. Because the interactions are rich, dynamic, part of a feedback loop and, above all, non-linear, the system's behaviour as a whole cannot be predicted by analysing its components.
- Complex systems are adaptive. They can reorganize their internal structure without requiring external intervention.

Transitions in societal systems, such as in food, water, energy, healthcare, mobility, or education, all involve complex adaptive systems (CAS). Complex systems are considered adaptive when their interconnected components can adapt and 'learn' from previous experiences and the system's surroundings (Holland 2006). Such complex adaptive systems (CAS) could be characterized as hierarchies of components interacting within and across scales with emergent properties that

4.5 Complex (Adaptive) Systems

Table 4.2 Illustrations of where natural design principles could be translated into various types of innovations that support a sustainable, healthy, and just society

Name	Natural design principle	Brief description	Type of (potential) innovation
Nature-based solutions to climate change adaptation	Resilience	An approach that uses biodiversity and ecosystem services to help people adapt to the adverse effects of climate change (Kabisch et al. 2017)	Climate change adaptation in rural and urban areas
Production of mushrooms on local residual flows	Circularity	Companies grow different kinds of mushrooms based on residual flows from other companies, and sell the mushrooms as a valuable new product to restaurants and others, while the substrate can be used for feed or other purposes.	Technological, Organizational
Species richness helps system respond to disturbance > healthy ecosystems demonstrate sustainability	Resilience, adaptive capacity, sustainability, circularity	Ecosystems survive biotic and abiotic disturbances by having multiple species that respond in different ways (Main 1999; Freudenstein et al. 2017). Likewise, diversity and life-span of plants help prairie ecosystems use water and nutrients efficiently (Glover et al. 2010)	Ecological, social, institutional
Adaptive management	Resilience and adaptive capacity	Natural ecosystems survive variable conditions, disease emergence, herbivores, predators, and other environmental changes by having adaptive responses (Tompkins and Adger 2004)	Institutional, social, organizational, community resilience, and examples of adaptive spatial planning
ManagemANT	Behaviourial strategies, self-organization	ManagemANT is a term which combines the words 'management' and 'ant', and describes the usage of behavioural strategies of ants in economic and management strategies (Fladerer and Kurzmann 2019).	Social, economic

(continued)

Table 4.2 (continued)

Name	Natural design principle	Brief description	Type of (potential) innovation
Social networking aids housing search	Networking	Social networking behaviours in hermit crabs help them find new homes through vacancy chains (Rotjan et al. 2010)	Social
Collaboration benefits multiple participants	Collaboration, networking	Several species of epiphytes, ants, fungi, and butterflies in mangrove forests provide benefits to each other through mutualism (Hogarth 2015).	Social, organizational
Community food forestry/Community agro-forestry	Resilience, circularity, sustainability, social cohesion	A food forest is a multi-functional approach to increase food security and provide ecosystem services (Clark and Nicholas 2013). A well-functioning food forest is actually the most ideal form of circular agriculture, since a healthy ecosystem shows an important basic principle of the Circular Economy: that is how you organize circularity at the lowest possible level. When combining a food forest with the concept of community-supported agriculture it provides many opportunities for social cohesion and citizen participation to restore the relationship between citizens, food production and nature. Community Food Forestry (CFF)/Community Agro-Forestry (CAF) allows citizens to become co-owner of their own food system. A first pilot is being established by	Ecologic, economic, and social

(continued)

Table 4.2 (continued)

Name	Natural design principle	Brief description	Type of (potential) innovation
		Inholland University in the Netherlands.	
Collaborative ecosystem/Creating value in business ecosystems	Collaboration, networking	Breakthrough business models can be created by companies working beyond the traditional supply chain, and create and scale system-level impact by working with non-traditional partners—e.g. competitors, other industries and sectors (e.g. Iansiti and Levien 2004; Clarysse et al. 2014).	Economic

cannot be predicted by knowing the components alone (Lansing 2003). As a consequence, 'an essential aspect of such systems is nonlinearity, leading to historical dependency and multiple possible outcomes of dynamics' (cf. Levin 1998).

Although complex adaptive systems can be hierarchical, they more often exhibit aspects of 'self-organization' (Holland 1995). Control is distributed rather than central (Allen and McGlade 1986; Pahl-Wostl 1995). Rather than trying to change the structure of complex, adaptive systems to make them controllable by external intervention, innovative management approaches aim at making use of the self-organizing properties of the systems to be managed. Ostrom (1990) convincingly shows that user communities of a common pool resource have the capacity for self-organization and self-governance and that there are many different viable combinations between the public and private sectors. Because the self-organizing properties of complex ecosystems and associated management systems seem to cause uncertainty to grow over time, understanding should be continuously updated and adjusted, and each management action viewed as an opportunity to further learn how to adapt to changing circumstances (Carpenter et al. 2001). The capacity to adapt to and shape change is an important component of resilience in a social-ecological system (Olsson et al. 2004; Lebel et al. 2006; Berkes and Turner 2006). In short, complex adaptive systems are characterized by self-organization, adaptation, heterogeneity across scales, and distributed control.

The difficulty in analysing complex adaptive systems is the multitude of relevant variables and the interactions of said variables, which all affect the functioning of the systems on multiple levels. This complexity only increases when social systems and natural systems are interlinked, as is the case in water management, agriculture, forestry, fishery, spatial planning, and environmental quality.

Our cities, for instance, are actually complex, adaptive, and multi-functional systems with a high degree of complexity and uncertainty. An important reason for this complexity is the dynamic interplay between a plethora of different systems in a single location. This goes for physical systems, such as buildings, streets, traffic systems, sewerage, gas, water, electricity, and communication networks, and the like, as well as non-physical systems, such as the job market, the economy, education, healthcare, politics and governance, law enforcement and, of course, the socio-cultural system at the city, district and street level. This also means that many different actors are involved in governing and managing complex systems, each of which has their own values, interests, ambitions, and opinions about problems and solutions. In other words, there are many interrelated factors and actors that can influence the functioning and governance of our cities and rural areas. Uncertainty is a given that cannot be brushed aside or ignored but must be addressed if we are to organize and design an agile, resilient society. Examples of the uncertainties faced by society include climate change, economic developments, financial crises, political shifts, disasters, new technologies, and diverging views on problems, solutions, and impact.

The characteristic features of complex adaptive systems, such as a high degree of variety and uncertainty, has a high risk of failure for any attempt at direct planning (Verhees 2013). Steering and coordination of complex adaptive systems, therefore, requires adaptive governance or reflexive governance, the latter being defined by Voß and Kemp (2015: 8) as: 'the organization (modulation) of recursive feedback relations between distributed steering activities'. This strategic process requires five key elements (Voß and Kemp 2015: 17–20): (1) transdisciplinary knowledge production; (2) experiments and adaptive strategies and institutions; (3) anticipation of long-term effects of measures; (4) interactive participatory goal formulation; and (5) interactive strategy development. Governance approaches that are capable of dealing with complexity and uncertainty in complex adaptive systems are highlighted in the following Sect. 4.6.

Complex system sciences provide valuable insights into the possibilities for transition in a complex system, such as the water, food or energy system, and the sustainability transition more in general. An important insight is that the outcomes or symptoms (events) of a system are determined by underlying behavioural patterns and interactions, whereby those patterns are linked to systemic structures such as biophysical conditions, markets and legislation, which in turn are again determined by mental models or paradigms such as beliefs, traditions, or (cultural) values (Maani and Cavana 2007; see Fig. 4.3). In particular, the mental models, the bottom layer in Fig. 4.3, determine or maintain the structures and decision-making of a system. Many scholars argue that the sustainability transition requires a paradigm shift. A paradigm shift implies a change in mental models and core values, though I doubt whether this is truly necessary for a substantial part of the core values already present in our modern societies and constitutions, such as freedom, equality, justice, and solidarity. Nevertheless, it is clear that the divide between humans and nature that arose during the Enlightenment, and the capitalist economic logic and related economic structures that were put in place after the Second World War, have blurred

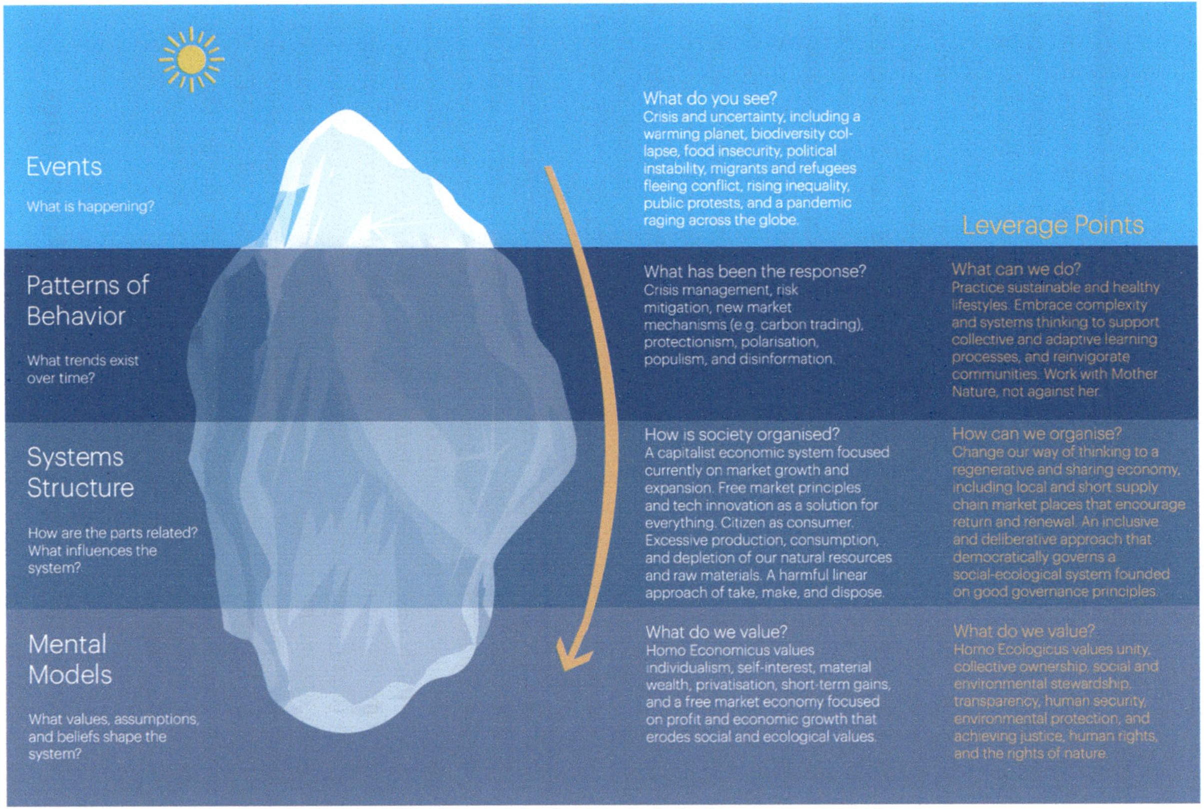

4.5 Complex (Adaptive) Systems

Events
What is happening?

What do you see?
Crisis and uncertainty, including a warming planet, biodiversity collapse, food insecurity, political instability, migrants and refugees fleeing conflict, rising inequality, public protests, and a pandemic raging across the globe.

Patterns of Behavior
What trends exist over time?

What has been the response?
Crisis management, risk mitigation, new market mechanisms (e.g. carbon trading), protectionism, polarisation, populism, and disinformation.

Systems Structure
How are the parts related? What influences the system?

How is society organised?
A capitalist economic system focused currently on market growth and expansion. Free market principles and tech innovation as a solution for everything. Citizen as consumer. Excessive production, consumption, and depletion of our natural resources and raw materials. A harmful linear approach of take, make, and dispose.

Mental Models
What values, assumptions, and beliefs shape the system?

What do we value?
Homo Economicus values individualism, self-interest, material wealth, privatisation, short-term gains, and a free market economy focused on profit and economic growth that erodes social and ecological values.

Leverage Points

What can we do?
Practice sustainable and healthy lifestyles. Embrace complexity and systems thinking to support collective and adaptive learning processes, and reinvigorate communities. Work with Mother Nature, not against her.

How can we organise?
Change our way of thinking to a regenerative and sharing economy, including local and short supply chain market places that encourage return and renewal. An inclusive and deliberative approach that democratically governs a social-ecological system founded on good governance principles.

What do we value?
Homo Ecologicus values unity, collective ownership, social and environmental stewardship, transparency, human security, environmental protection, and achieving justice, human rights, and the rights of nature.

Fig. 4.3 The iceberg model: four levels of a system (based on Meadows (2008), Maani and Cavana (2007) and others) and adapted by author to reflect a societal transformation towards a Natural Social Contract

or ignored a number of important core values, such as social and environmental stewardship, planetary health, environmental security and justice, intergenerational justice and equity, and the Rights of Nature (see Sect. 3.8). Hence, it may be more important to resurrect the core values that are latent or have been blurred by current systemic structures. For example, during the Corona-pandemic we have witnessed a resurrection of solidarity, and a re-appreciation of people working in health care, food production, education, who were regarded as mere production factors in a neoliberal model and who had to hold up their hands for a decent salary before the Corona-crisis. They are now the heroes of society. Of course they already were, but appreciation for these professionals failed to materialize because it was overshadowed by the over-appreciation of the free market, market-based values, privatization, and unlimited economic growth. The point here is that current day values are also influenced by the systemic structures that we have put in place, due to a process of institutionalization, in which preferences are gradually strengthened until they are fixed and familiar (Parker et al. 2003). Hence, it is important to realize that the divide between humans and nature, as well as the capitalist economic logic, has gradually entered present day constitution, laws and legislation, as well as customary law, existing practices, norms and culture. However, there is little societal and scientific attention for the underlying patterns, structures, and paradigms that cause the symptoms (the events in Fig. 4.3) of systemic failures to recur over and over (IPBES 2019; Wallace et al. 2015). From this perspective, global warming, loss of biodiversity, and environmental degradation are merely symptoms of a deeper, systemic crisis, but because they are very visible they receive a lot of (political) attention.

An important question is where leverage points in a complex system can be found. These leverage points are places in a complex system where a small change could bring about major changes (Meadows 2008). As these leverage points focus on the deeper layers of the system, they increase in impact and are more transformative on the system (both positive and negative). Finding leverage points alone is not enough; system change also requires good insight into the interrelationships, for example, via (non-linear) feedback loops, and how the desired outcome can be achieved with maximum synergy effects and minimal 'trade-offs' (Kennedy et al. 2018). Adopting a systems-based approach helps recognize synergies and trade-offs, moving beyond linear, to more circular, inclusive systems' (cf. SAPEA 2020). 'Systems thinking is about seeing life in motion, recognizing that the big picture is rarely static, but almost always a web of factors that interact to create patterns and change over time' (Martella et al. 2019).

4.6 Adaptive, Reflexive, and Deliberative Approaches to Governance

The complexity and uncertainty of the problems we face today call for new forms of governance, management, and organization. In particular for sustainability development there is a broad and diverse field of governance studies that propose **adaptive**

4.6 Adaptive, Reflexive, and Deliberative Approaches to Governance

governance (e.g. Folke et al. 2005; Huitema et al. 2009; Termeer et al. 2010; Huntjens et al. 2011a, b; Ison et al. 2013; Chaffin et al. 2014), **reflexive governance** (e.g. Rip et al. 2006; Leach et al. 2007; Hendriks and Grin 2007; Voß and Bornemann 2011; Voß and Kemp 2015; Feindt and Weiland 2018), and **deliberative governance** as new pathways to sustainability. These concepts share a focus on addressing ambivalence, complexity, uncertainty, and distributed power in societal change. These governance concepts have been translated into various management approaches, of which transition management (as described in Sect. 4.2) and adaptive management (in this section) are two examples that evolved from the analysis of sociotechnical systems and social-ecological systems, respectively (Voß and Bornemann 2011).

The concept of adaptive management has been known for longer, especially within ecosystem management (Holling 1978; Walters 1986; Pahl-Wostl 1995; Lee 1999). An important starting point is that ecosystems are complex systems that are adaptive or self-organizing, and that management systems must, therefore, be able to respond to changes or surprises in the system (Gunderson and Holling 2002). The ability to adapt is an important part of resilience in a social-ecological system and is also called adaptive capacity (Berkes et al. 2002; Folke et al. 2005; Walker et al. 2004). Adaptive management is a systematic process to improve policy and practice by learning from the outcomes of previous activities and by taking into account changes in external factors, which is also known as a 'management as learning' approach (Gunderson et al. 1995).

The application of adaptive management has far-reaching consequences for policy and strategy development, and the translation of adaptive management into policy and strategy is what we call adaptive planning or adaptive governance. Adaptive planning requires strategies that can be adjusted in time (in the event of changing circumstances), space for experimentation and deliberation of alternative routes and measures (Huntjens et al. 2012). Ahern (2006) emphasizes that a transdisciplinary process, in which a certain level of uncertainty and risk is accepted, is necessary to achieve adaptive planning, which is why it is so important to activate a social learning process geared towards the process rather than a fixed goal (Bagheri and Hjorth 2007).

For example, there is a real risk that current climate and energy policies will be too rigid to achieve the policy objectives for carbon emission reduction. Because this policy is linked to an energy transition that is set to last several decades, it is inevitable that, a few years from now, we will find that a number of things turned out differently from how we thought they would. A more adaptive and robust policy should, on the one hand, provide investment security for businesses and, on the other hand, offer sufficient scope for adjustment in the light of new insights and political, economic, social, and technological developments. As things are now, for example, we cannot predict exactly which role hydrogen will play in the energy supply of the future, how quickly mobility will become electric, or how much progress will be made in the field of nuclear energy. Similarly, the consequences of new technologies such as artificial intelligence, quantum computers, blockchain technology, and big data are still unknown. It is, therefore, important to ensure that we do not develop climate and energy policies of such rigidity that they cannot be adjusted at a later stage.

In general, adaptive management and transition management provide a number of important insights for Transformative Social-Ecological Innovation, of which the importance of collective learning processes and social networks for coping with uncertainty and enabling change stands out. Collective learning processes (see Sect. 4.7) are required to develop the knowledge and ability to respond to new insights, with mechanisms that facilitate social learning, policy learning, and transformational learning (Huntjens et al. 2012; Beers et al. 2016). For instance, a transdisciplinary approach is required to enable collective learning (Sect. 4.10). Beyond interdisciplinary cooperation, a transdisciplinary cooperation between citizens, businesses, government, and other parties stimulates creativity, generates support for solutions, and allows for the production, exchange, and use of practical knowledge (see Sect. 4.10).

4.7 Social Learning, Policy Learning, and Transformational Learning[1]

Society's capacity to learn is perhaps the most essential property for realizing a societal transformation towards a Natural Social Contract. There is a large and diverse body of literature on various forms of collective learning, including literature on action learning, social learning, organizational learning, policy learning, and transformational learning. **Action learning** offers a widely accepted framework for understanding and engaging systematically in practical knowledge construction (Levy 2003). The concept of action learning is being applied in organizational learning (Argyris and Schön 1978, 1996), business management (Sterman 2000), financial sector, health sector (Levy 2003, on community empowerment; Hanks 2006, on community partnership), educational sector (Hwang 2000; Maurer et al. 2006), as well as in the agricultural sector, water management (2007; Huntjens et al. 2011a, b), and social innovation in sustainability transitions (Huntjens 2019). **Social learning** means learning together to solve a collective problem (Craps 2003; Pahl-Wostl 2007). Social learning happens when people with different goals and resources successfully tackle a problem in which all have a stake (Craps 2003). **Policy learning** is an important concept in the field of public administration (Hall 1988; Bennett and Howlett 1992; Sanderson 2002; Kemp and Weehuizen 2005; Leicester 2007; Grin and Loeber 2007; Sabatier 1988; Sabatier and Jenkins-Smith 1993; van Buuren et al. 2016). Policy learning is defined as a 'deliberate attempt to adjust the goals or techniques of policy in the light of the consequences of past policy and new information so as to better attain the ultimate objects of governance' (Hall 1988, 6). Policy learning involves a socially conditioned discursive or argumentative process of developing cognitive schemes or frames that question policy goals and

---

[1]Parts of this section are based on Huntjens (2011).

assumptions (Sanderson 2002, 6). **Transformational learning** (also called triple loop learning) is a refinement of the original double loop learning concept by Argyris (1999), and helps to bring about fundamental shifts in thinking and attitude (Hargrove 2002:60). It starts with declaring powerful new possibilities for governance and management and then translating them into goals that take people and organizations beyond what they already think and know based on their own or organizational orthodoxies or experience (Hargrove 2002: 115), or to take them beyond their old management styles.

Lave and Wenger (1991) emphasize that a collective learning process cannot be divorced from the social context in which it takes place. At the core of this social theory of learning lies the concept of a community of practice (ibid.). This means that individuals learn by taking part in practice and gradually shift more to the core of the process, but the practice also participates in individuals by influencing thoughts and actions. This continuous change of perspective is particularly effective at highlighting the social dynamics and context of learning, with learning being considered a part of everyday life rather than a process that takes place solely in people's heads (ibid.).

An important hypothesis advanced in the literature is that cooperation with stakeholders, starting as early as possible in the process, promotes collective (or societal) learning (Boonstra 2004; Hisschemöller 2005; Muro and Jeffrey 2008; Stringer et al. 2006). It helps generate trust, develop a shared understanding of problems, solve conflicts and find shared solutions, ultimately enabling all stakeholders to achieve better results than they would have on their own (Craps 2003). Beers et al. (2016) address the relationship between social learning and transitions, empirically showing how the nature of the interaction relates to the outcomes and impacts of social learning.

Policy changes have been explained in terms of learning by Sabatier and Jenkins-Smith (1999) through their Advocacy Coalition Framework (ACF). However, one limitation in the ACF is that advocacy coalitions take their identity from core beliefs, they are conservative of them and thus also of the policy positions they advocate (Weible et al. 2009). Conservatism leads Sabatier and Jenkins-Smith (1993) to propose that collective learning appears not from change within policy coalitions, but as a result of the changing influence of policy coalitions on the whole. In this model, the system learns without any learning on the part of policy coalitions or individuals. Movement is argued to be stimulated by shocks and trends exogenous to the system—including wider political change, legislative reform, or stressors such as climate change. By doing so, the ACF does not explicitly account for, or is ambiguous about, the role of ideas and self-interest in the policy process (e.g. Kübler 1999; Compston and Madsen 2001).

As Argyris and Schön (1996) have shown, changing values is far more difficult than changing practices. Argyris and Schön consider double loop learning more difficult than single loop learning, because it requires changes to values. Individuals tend to avoid challenging established values. Argyris and Schön (1996) argue this is for three reasons:

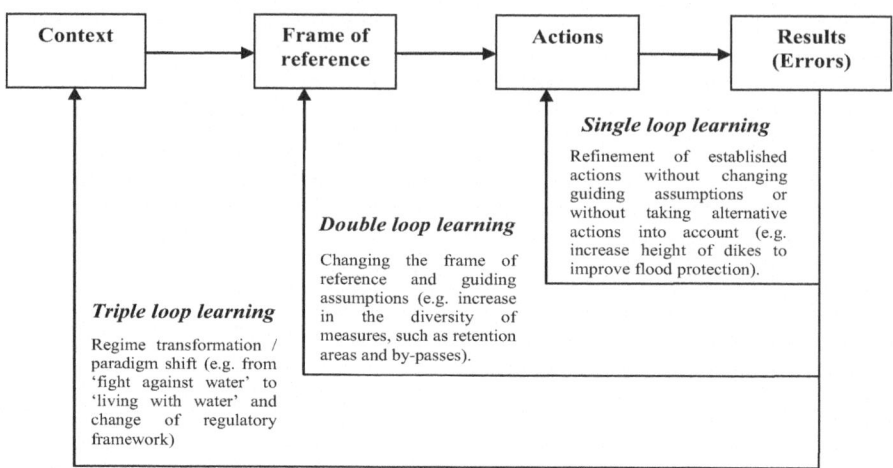

Fig. 4.4 Triple loop learning concept derived from Hargrove (2002), and adjusted by Huntjens et al. (2011a, b)

- Individual risk aversion that leads actors to avoid direct interpersonal confrontations and public discussion of sensitive issues which might expose the actor to future negative repercussions.
- A desire to protect others by avoiding the testing of assumptions where this might evoke negative feelings and by keeping others from exposure to blame.
- A wish to control the situation by keeping your own view private and avoiding any public questioning which might refute it.

It is important to recognize that policies change in a variety of different ways. As has long been recognized, some policies are new and innovative, while others are merely incremental refinements of earlier policies (Hogwood and Peters 1983; Polsby 1984). In other words, policy learning may have different levels of intensity (Pahl-Wostl et al. 2007). Some scholars have conceptualized societal transformation as social and societal learning that proceeds in a stepwise fashion moving from single to double to triple loop learning (Pahl-Wostl 2009; Huntjens et al. 2011a, b), making use of the concept of double loop learning (Argyris 1999) and triple loop learning (Hargrove 2002), as an extension of the double loop concept (see Fig. 4.4).

Section 5.3 provides an operationalization and framework for analysing different levels of collective learning in societal transformation. Section 5.4 will provide more detail on collaborative action research, and related methods to stimulate and facilitate interaction, participation, social learning, and co-creation.

4.8 Shared Value, Multiple Value Creation, and Mutual Gains

For studying Transformative Social-Ecological Innovation (TSEI) it is necessary to pay attention to the incentives, thinking, and actions of the actors involved and the conditions that influence this. Special attention is paid to the values that the actors use and the way in which they are expressed in the cooperation. There are various approaches that provide useful insights, such as shared value creation (Porter and Kramer 2002, 2019), integrated value creation (Visser and Kymal 2015), mutual gains approach (Susskind and Field 1996; Susskind and Cruikshank 2006; Rodríguez-Carvajal et al. 2010; Ryan and Wallace 2019), multiple value creation, and co-creation (see Sect. 4.8). In this section I will briefly highlight some key insights from this literature.

Porter and Kramer (2002, 2019), Porter et al. (2012) introduced the concept of Creating Shared Value (CSV), which is closely related to the concept of Corporate Social Responsibility (CSR) (De Witte and Jonker 2006). Porter and Kramer make a strong plea for a fundamental revision of capitalism, although still very much relying on market mechanisms, profit making (although broader defined), and the 'competitive context' of society. Their argument is that by serving social and ecological interests, companies will earn money. After all, companies are focused on innovation and with that they excel the change towards a better society. According to Porter and Kramer, capitalism remains the best and most efficient way to create value, but traditional capitalism sees only 'profit' as value. The government and other powers should try to ensure that this pursuit of profit does not lead to serious abuses. The new capitalism of Porter and Kramer pays much more attention to the common interest and creates value based on social interest and not only on the interest of the shareholders. Society has an enormous amount of needs that are not or insufficiently met. Think, for example, of solving issues in the field of the environment, health care, the scarcity of raw materials, education, and employment.

According to Porter et al. (2012), companies can create shared value opportunities in three ways:

1. Reconceive products and markets to provide appropriate services and meet unmet needs.
2. Redefine productivity in the value chain to mitigate risks and boost productivity.
3. Enable local cluster development by improving the external framework that supports the company's operations, for example, by developing the skills of suppliers.

Moon et al. (2011) added one more step to define core competence, while also incorporating internationalization, an aspect that was missing in Porter & Kramer's work:

4. Enabling local or global cluster development (Moon et al. 2011).

LEVELS OF SHARED VALUE	BUSINESS RESULTS	SOCIAL RESULTS
Reconceiving product and markets: How targeting unmet needs drives incremental revenue and profits	• Increased revenue • Increased market share • Increased market growth • Improved profitability	• Improved patient care • Reduced carbon footprint • Improved nutrition • Improved education
Redefining productivity in the value chain: How better management of internal operations increases productivity and reduces risks	• Improved productivity • Reduced logistical and operating costs • Secured supply • Improved quality • Improved profitability	• Reduced energy use • Reduced water use • Reduced raw materials • Improved job skills • Improved employee incomes
Enabling cluster development: How changing societal conditions outside the company unleashes new growth and productivity gains	• Reduced costs • Secured supply • Improved distribution infrastructure • Improved workforce access • Improved profitability	• Improved education • Increased job creation • Improved health • Improved incomes

Fig. 4.5 Levels of shared value creation and types of value created in each area (Porter et al. 2012)

Porter et al. (2012) provide further examples of the types of value created in each area (see Fig. 4.5).

A critical study on shared value creation by Pirson (2012) found that 'innovative shared value creating ventures opted out of balance-oriented, shared value creation strategies and embraced either financial or social-value primacy strategies over time. The findings thus question the power of the shared value creation notion when viewed as balance orientation' (cf. Pirson 2012). A common criticism of CSV is the downplay of trade-offs that businesses have to make (The Economist 2011). Furthermore, a literature review on shared value creation by Williams and Hayes (2013) shows there is little documentation of its influence elsewhere, with exception of some examples, with varying degrees of success, of US-based multinationals. 'There has been little rigorous analysis into the impact of CSV mechanisms, with the majority of evidence existing as standalone case studies of mixed analytical rigour' (cf. Williams and Hayes 2013). London (2009) furthermore argues that the predominant focus in terms of social impact is on income, missing wider social-ecological dimensions. 'All current measurement models suffer from standard impact challenges, with the emphasis on tasks completed or products distributed rather than outcomes' (cf. Williams and Hayes 2013).

In addition to shared value thinking there is literature on the concept of integrated value creation (IVC), which resembles shared value creation, but is not exactly the same (Visser and Kymal (2015). Practically, 'IVC helps a company integrate its response to stakeholder expectations (using materiality analysis) through its management systems (using best governance practices) and value chain linkages (using life cycle thinking)' (cf. Visser and Kymal 2015). Visser (2017) provides the following working definition of IVC: "integrated value is the simultaneous building of multiple 'non-financial' capitals (notably infrastructural, technological, social, ecological, and human capital) through synergistic innovation across the nexus economy (including the resilience, exponential, access, circular, and well-being economies) that result in net-positive effects, thus making our world more secure, smart, shared, sustainable, and satisfying".

Originating from political sciences, and negotiation theory more in particular, the mutual gains approach (MGA) offers valuable insights for complex multi-party problem solving. It has been successfully used in many negotiations, mainly related to trade, labour, and environmental negotiations (Susskind and Field 1996; Kirk et al. 2008; Rodríguez-Carvajal et al. 2010; Ryan and Wallace 2019), while also applied in citizen engagement, process facilitation, mediation, and conflict resolution, for instance, to mediate in the Israeli–Palestinian water conflict (Huntjens 2017).

The mutual gains approach is highly valuable in situations where two or more people are negotiating to reach an agreement that may be of benefit to both or all of them (Consensus Building Institute 2014). The MGA-approach lays out four steps for negotiating better outcomes while protecting relationships and reputation. The 4 phases of the mutual gains process include (cf. CBI 2014):

1. **Preparation**: Prepare by understanding interests and alternatives. More specifically, estimate your BATNA and how other parties see theirs (BATNA stands for 'best alternative to a negotiated agreement'). Having a good alternative to agreement increases your power at the table (Raiffa 1982; Fisher et al. 1991; Zartman and Rubin 2000).
2. **Creating Value**: Create value by inventing without committing. A central feature of the mutual gains approach is a focus on interests, not positions. Based on the interests uncovered or shared, parties should declare a period of 'inventing without committing' during which they advance options by asking 'what if...?' By doing so, parties can discover additional interests, create options that had not previously been imagined, and generate opportunities for joint gain by trading across issues they value differently (Fisher et al. 1991; Bazerman and Neal 1992).
3. **Distributing Value**: At some point in a negotiation, parties have to decide on a final agreement. This is easier to do when there is trust between the parties, and the more value they have created, the easier this will be (Fisher et al. 1991), but research suggests that parties default very easily into positional bargaining when they try to finalize details of agreements (Mnookin et al. 2000). Parties should divide value by finding objective criteria that all parties can use to justify their 'fair share' of the value created. By identifying criteria or principles that support or guide difficult allocation decisions, parties at the negotiating table can help the groups or organizations they represent to understand why the final package is not only supportable, but fundamentally 'fair'. This improves the stability of agreements, increases the chances of effective implementation, and protects relationships.
4. **Implementation and follow-up**: Follow through by imagining future challenges and their solutions. Parties near the end of difficult negotiations—or those who

will 'hand off' the agreement to others for implementation—often forget to strengthen the agreement by imagining the kinds of things that could derail it or produce future conflicts or uncertainty (Bazerman and Watkins 2004; Susskind and Cruikshank 2006). While it is difficult to focus on potential future challenges, it is wise to include specific provisions in the final document that focus on monitoring the status of commitments; communicating regularly; resolving conflicts or confusions that arise; aligning incentives and resources with the commitments required; and helping other parties who may become a de facto part of implementing the agreement (Lax and Sebenius 2006). Including these provisions makes the agreement more robust and greatly assists the parties who will have to live with it and by it (Susskind and Cruikshank 2006).

In the search for mutual gains, participants are encouraged to explore more ways to create more value (i.e. to increase the pie) and generate a broader vision on sharing benefits. To illustrate, whenever action is taken to remedy environmental problems, the benefits also cascade: for instance, nurturing wildlife and flora in a wetland can also reduce water pollution and soil erosion, and protect crops against storm damage, alleviating water scarcity and allowing for more food production. In other words, working on one aspect of human security (i.e. environmental security in above example) may contribute to other aspects of human security (i.e. water and food security in above example). During mediation in the Israeli–Palestinian water conflict this aspect of multiple value creation was demonstrated by a multi-functional usage approach, in which the same cubic metre of water is being used by multiple users at different points in time before it flows into the river, among other thanks to centralized or distributed waste water treatment and recycling (Huntjens 2017). A central tenet of the MGA-approach is that a vast majority of negotiations in the real world involve parties who have more than one goal or concern in mind and more than one issue that can be addressed in the agreement they reach. The MGA-approach allows parties to improve their chances of creating an agreement superior to existing alternatives.

4.9 Effective Cooperation

There is currently no academic consensus on the definition of 'effective cooperation', with some authors suggesting that cooperation is effective if it leads to mutual satisfaction between the parties involved (Grey et al. 2009, p. 19). However, mutual satisfaction need not necessarily coincide with effective cooperation, as satisfaction is a state of mind whereas effective cooperation often boils down to a combination of economic profits and political gains. Huntjens et al. (2016), therefore, state that cooperation can only be considered effective if, on the one hand, there is sufficient trust between stakeholders with different or even conflicting interests, in order to reach a mutually accepted agreement and, on the other hand, when the intended results of the agreement are achieved. Effective cooperation can, therefore, be defined as 'cooperation in which two or more parties come to a negotiated compromise on maximizing mutual benefits and achieving shared gains for the parties involved' (cf. Grey et al. 2010, p. 158).

Maximizing mutual benefits, however, is only one of the factors that contribute to mutual satisfaction. For example, focusing on effectiveness can turn out for the worse (for examples, see New Public Management, Dunleavy et al. 2006) if results become more important than the process by which they are achieved. As such, procedural justice and psychological satisfaction are two other important elements that lead to mutual satisfaction for all parties involved (Creighton et al. 1998, p. 55; Lawrence et al. 1997).

In order for cooperation to be effective, it is important that the cooperating and decision-making process is conducted in a fair and equitable manner (see the literature about procedural justice and distributive justice). The solution is often to bring stakeholders together at an early stage, so that they can all contribute their ideas to the process, cooperate, and take part in decisions. This ensures that solutions enjoy more support and are better adapted to their social context, boosting their effectiveness.

It goes without saying that simply reaching an agreement is not enough: the intended results must also be achieved. If the parties involved fail to achieve what they set out to do, the process cannot be considered effective cooperation and stakeholders will be dissatisfied. Achieving the intended results is also called the level of compliance, which is used as a measure of institutional effectiveness (Biermann et al. 2007, p. 10).

Multi-party cooperation is paramount for issues and developments:

- that matter to multiple stakeholders.
- that involve various stakeholders who depend on each other to achieve their goals.
- that are characterized by incomplete or distributed knowledge.
- where there is little consensus about the problems at hand or the solutions to these problems.

Table 4.3 provides an overview of perceived advantages and disadvantages of cooperation.

4.10 Transdisciplinary Approach, Living Labs, and Citizen Science

The sustainability transition requires new forms of cooperation, organization, and governance, which calls for the commitment and creativity of many different people who work and live in the district, city, or region in question, as well as people working on urban and regional development. This, in turn, will necessitate more parties, such as real estate parties, project developers, municipalities, provinces and ministries, water boards, SMEs, farmer cooperatives, citizens, and NGO's, as well as scientists from different disciplines to start working together.

Table 4.3 Perceived advantages and disadvantages of cooperation

Possible advantages of cooperation	Possible disadvantages of cooperation
• win-win solutions as a result of broadening the scope, • new knowledge through social learning, • locally adapted solutions, • more support for solutions, • prevention of claims and litigation, • improved communication and cooperation between people and organizations, • mutual trust between people, organizations, and authorities, • improved social cohesion, • shared ownership and buy-in, • scope for different needs and interests, • improved funding opportunities,	• time-consuming, • decisions tend to be compromises, • disappointment because of unrealistic expectations, • project process can be chaotic and difficult to plan and predict, • solutions that may not technically be the best solutions to the problem, • (partial) loss of control • process management/facilitator (rather than only project implementation), • higher costs (e.g. for extra facilities, travel expenses for participants and professional facilitators), • liability/risk for not involving the right or all stakeholders, or the process is perceived as biased by certain shareholders.

Combining knowledge from different disciplines is known as multidisciplinarity, while cross-field cooperation between academic disciplines is termed interdisciplinarity. However, solely integrating knowledge from different fields is not enough to tackle complex societal issues. Rather, this must be combined with a participatory approach, resulting in what is called transdisciplinarity (Van Buuren en Edelenbos 2004; Reed 2008; Pahl-Wostl 2007). Transdisciplinary research involves interdisciplinary academic cooperation in collaboration with societal actors who do not necessarily have an academic background, with the aim of developing knowledge that is relevant for practice.

In particular, there is a need for such a transdisciplinary approach when there is incomplete knowledge about an issue, when knowledge is distributed among multiple stakeholders, and when there is little consensus about the problems at hand or the solutions to these problems. By definition, TSEI processes are marked by a high degree of complexity and uncertainty, as they often involve a multitude of interrelated factors and actors. As such, there is more to such processes than technological innovations and interventions alone, and social, cultural, economic, and administrative aspects also play a role, as does the process of multi-party cooperation.

In this respect, it is important not only to facilitate interdisciplinary cooperation but also, and above all, to involve citizens, entrepreneurs, governments, and other practically relevant parties (transdisciplinarity). By doing so, it becomes possible to produce, exchange, and use practically relevant knowledge. A transdisciplinary approach is needed to mobilize system knowledge, promote creativity, and generate support for new solutions (see also Wicked Philosophy by Coyan Tromp 2018).

In practice, a transdisciplinary approach for addressing sustainability issues has found its way in various countries in the form of living labs, which on itself could be considered a societal innovation. Living labs are physical locations where fundamental, applied and practice-oriented researchers, citizens, businesses, and government agencies work together to solve societal problems in a lifelike setting. A living lab is an open innovation ecosystem in which products or services are developed, tested, and used together with the people, businesses, or organizations who will actually use them. This real-life environment is essential for the development of innovative solutions that can survive the complexity of real life and daily practice (Maas et al. 2017). Often, organizational and/or societal innovations turn out to be decisive (ibid.). The setting of a living lab allows participants to study various aspects of physical and social systems and the relationships between such systems, focusing in particular on human interaction with systems and cooperation between stakeholders. This makes living labs a valuable component of the research methodologies adopted by this research group.

Within the broad field of transdisciplinary approaches, citizen science is becoming increasingly popular. Citizen science is sometimes described as 'public participation in scientific research', and often referred to in relation to participatory monitoring and participatory action research (Irwin 1995; Hand 2010; Bonney et al. 2014; Doyle et al. 2019). For example, citizen science is actively used for biodiversity monitoring (Theobald et al. 2015; Chandler et al. 2017), for monitoring of (micro)plastics and their associated pollution (Turns 2019), for crop variety selection for climate adaptation, involving thousands of farmers (Van Etten et al. 2019), and more recently, also for monitoring the spread of the Coronavirus (Covid-19) pandemic. As such, citizen science provides an important avenue for the co-creation of actionable knowledge and solutions (Santha 2020). In particular, Internet and smartphones have increased the options for citizen science. Section 5.4 provides more detail on action research, and collaborative action research in particular, which refers to the involvement of practitioners and stakeholders in practice-driven research, instead of only citizens and scientists, which is usually the case in citizen science.

4.11 The Art of Co-creation: Approaches, Principles, and Pitfalls

At its core, Transformative Social-Ecological Innovation (TSEI) requires a process of co-creation tailormade to purpose and context, in which different parties work together to solve a complex societal problem or challenge. But co-creation does not happen automatically, and requires a well-considered approach depending on many factors and accounts for complexity and uncertainty. In practice, numerous multi-stakeholder workshops and dialogues, design sessions, round table discussions, and brainstorming sessions along with hackathons, living labs, urban labs, and field labs have become popular ideation settings. And yet, it remains understudied how the co-creation of knowledge and practices develop and flourish within such multi-actor learning environments (Puerari et al. 2018). Most studies focus on the identification of influential factors, while hardly any attention is paid to the outcomes (Voorberg et al. 2015). Effectiveness of co-creation trajectories clearly depends not just on the quality of participation and facilitation, as is widely acknowledged, but also on the preparation of conveners and the follow-up actions of participants around main events (Huntjens et al. 2017). This raises an important question as to what extent the right method or approach has been chosen to actually realize co-creation, social learning, effective cooperation, and mutual trust. For every phase and aspect of a co-creation trajectory there exists a wide choice of approaches (see Table 4.4 below) and methods (see examples in Sect. 4.7) available to convene parties, inspire collective problem solving, achieve consensus or at least consent, and encourage effective implementation. In this section, therefore, I give an overview of different approaches from which I will distill some basic principles and pitfalls for co-creation.

Table 4.4 provides an illustrative overview (non-exhaustive and in no particular order) of the various approaches available for co-creation. Differences between these approaches depend on the context in which they are applied. For example, focus and purpose (e.g. open innovation, new business model, or systemic innovation), level of complexity in the area of application, methodology (including time and resources required), and the level of education and competences of participants and facilitators. Hence, choosing the most suitable approach determines the success of co-creation in terms of effectiveness, efficiency, and fairness (i.e. procedural justice). This requires considering several additional factors, such as:

- Nature of the problem or the challenge (e.g. field of application, level of complexity, shared sense of urgency, degree of diversity (or conflict) in interests and perspectives, expected outputs, outcomes, impacts, etc.)
- Intended goal or purpose of the co-creation process (e.g. product, service, agreement, plan, strategy, tactics, learning goals, etc.)
- Time and resources available.
- Competences of participants and facilitators regarding methodology, process, and content. This also includes openness to dialogue, and willingness to be explicit about one's underlying assumptions and mental models, etc.).
- Commitment and ownership of contributors/problem owners, in particular to what extent (intended) participants are willing to cooperate, contribute, share knowledge, and follow-up and implement the agreed upon actions or results of the co-creation trajectory. The latter relates to strategic knowledge management and how intellectual property rights (IPRs) can facilitate the sharing of technology and of know-how, thus supporting collaborative innovation.
- Safe environment for open dialogue and exchange (e.g. Chatham House Rules).
- Path dependence (history, possible tensions/conflicts, social relationships between participants, phase in planning or policy cycle, etc.)
- Ethical issues, for example, equal access to information generated by the process for all participants, a process that maximizes the opportunities for involvement of all participants (i.e. procedural justice), or responsibility for maintaining confidentiality (e.g. in case of Chatham House rules).

When a company or organization uses an open innovation model it recognizes that knowledge from multiple external sources is necessary to enhance innovation

Table 4.4 Overview of different approaches for co-creation (non-exhaustive and in no particular order)

Approach	Focus	Context
Open innovation model	Business model	Open innovation model is the mainstream model used for (mostly) technological innovations in open market economy (supply-side driven) (e.g. Chesbrough 2006; Lee et al. 2010; Gassmann et al. 2010)
Boundary work	Business model innovation for sustainability	Multiple value creation via cross-sector collaboration, which requires changes in the boundaries of identity, power, competence, and efficiency[a] (e.g. Zietsma and Lawrence 2010; Clark et al. 2016; Velter et al. 2020)
Design thinking	User-centred	Open market economy (demand-side driven) (e.g. Rowe 1987; Brown 2008; Dorst 2011; Black et al. 2019)
Shared value creation (SVC) /integrated value creation (IVC)	Business model/multiple value creation	Open market economy (SVC: Porter and Kramer 2002, 2019; IVC: Visser and Kymal 2015)
Mutual gains approach (MGA)	Conflict resolution/ multiple value creation / stakeholder management	Negotiation, mediation, and value creation in multiple stakeholder settings (e.g. international trade or environmental negotiations) (e.g. Susskind and Field 1996; Susskind and Cruikshank 2006; Rodríguez-Carvajal et al. 2010; Huntjens 2017; Ryan and Wallace 2019)
Reflexive interactive design (RIO-approach)	Systemic innovation	Transition management (e.g. Bos et al. 2009; Bos and Grin 2012; Bremmer and Bos 2017; Puente-Rodríguez et al. 2019)
Strategic niche management (SNM)/ transition experiments	Systemic innovation	Transition management (e.g. Kemp et al. 1998; Schot and Geels 2008; Raven et al. 2010; Witkamp et al. 2011; Luederitz et al. 2017)
Collaborative action research (CAR)/ Participatory action research (PAR)	Participatory knowledge development/ transdisciplinary collaboration	Wicked problems (e.g. Eden and Huxham 1996; Checkland and Holwell 1998; Baum et al. 2006; Reason and Bradbury 2005; Huntjens et al. 2011b, 2014a, b, c)

[a]Efficiency refers to the (perceived) efficient locus of transaction governance (Santos and Eisenhardt 2005)

and deliver additional value for customers or clients. This sits in contrast to closed innovation in which a company strives to generate the best ideas entirely on its own. Although the open innovation model purposefully uses the inflows and outflows of knowledge to accelerate innovation (Chesbrough 2006), it is by definition an internally oriented process with a prime focus on maximizing utility and market expansion to the benefit of the company itself. Other approaches, such as shared value creation (SVC) and integrated value creation (IVC) stress the common interest with the purpose to create value based on social and ecological interests and not only on the interest of the shareholders (see Sect. 4.8).

SVC and IVC approaches argue that by serving social and ecological interests, companies will earn money. As such these approaches rely heavily on market mechanisms, profit making (although broader defined), and the 'competitive context' of society, and, therefore, tailored to an open market economy. Other approaches, such as the mutual gains approach (MGA) or those more related to transition management (e.g. RIO and SNM), focus explicitly on solving complex or wicked societal problems or challenges as something that requires changes in dominant structures (regimes). This is particularly true where multiple parties recognize mutual dependency and the importance of finding common ground and values in order to arrive at collective problem solving, systemic innovation, and joint implementation. These approaches have been explicitly developed to better deal with complexity and uncertainties with specific attention for issues of politics and power and drawing on the wider field of governance and innovation studies as well as other fields like complexity theory and systems theory.

Despite these differences in approaches to co-creation there remain many common denominators in the process of co-creation. Based on a literature review (e.g. Voorberg et al. 2015; Huntjens et al. 2017; Puerari et al. 2018) and drawing from years of practical experience in many co-creation trajectories—either as convener, mediator, organizer, or participant—I have identified a number of basic principles and common pitfalls of co-creation (Table 4.5).

4.11 The Art of Co-creation: Approaches, Principles, and Pitfalls

Table 4.5 Principles and pitfalls of co-creation

Principle of co-creation	Explanation and potential pitfalls
Focus on collective problem solving	Co-creation involves collective problem solving with multiple parties, and where multiple parties recognize mutual dependency and the importance of finding common ground, shared values, and mutually accepted solutions. Dominance of one-sided interests or power, hidden agendas, lack of trust, incomplete configuration of problem owners and knowledge providers, and biased or incomplete information are common pitfalls. Likewise, a lack of commitment, ownership, and enthusiasm are common obstacles to co-creation, often due to an absence of attractive value propositions for all.
Safe environment	A safe environment for open, empathetic, and equal dialogue, and open exchange of information and reciprocity (e.g. through Chatham House rules) are important requirements for co-creation. It is supported by equality and trust among participants. e.g. by using a Round Robin approach where every participant has an equal voice and opportunity to co-create. Power imbalances and mistrust between participants may feed feelings of insecurity and exacerbate an unwillingness to co-create. In this vein, insufficient insights into complex social relationships and history between participants risks creating a weak environment for exchange and reciprocity.
Skilled and neutral facilitation	Co-creation requires a structured, safe, and creative process under the guidance of a skilled and neutral process manager, facilitator, or coach. Complex trajectories of co-creation usually involve more people in various roles to organize and facilitate the process. An important factor is the right choice of approach and methods, whereas more advanced approaches or methods (e.g. for one step or phase in the co-creation process) may require ample experience and skills. Also, the neutrality of the facilitator needs to be accepted by all parties and requires mediation skills in case of conflicting parties.
Delicate interplay between process and content	Co-creation relies on a well-informed and comprehensive understanding of the collective problem, problem owners, different interests and perspectives, and possible solutions which can be achieved by open, shared, and multiple information sources that fill knowledge gaps and facilitate integration. At the same time, it needs to be recognized that many uncertainties cannot be resolved. A multi-disciplinary advisory board could help to overcome knowledge gaps and other barriers to creativity. Too much emphasis on action and results could lead to a lack of reflection, superficial treatment of problems, and a lack of scientific support. Vice versa, too much emphasis on a science-proof co-creation process could serve as a limitation for creativity and out-of-the-box thinking. Depending on the subject and the different parties, expertise, and disciplines involved in the process, communication and language problems regularly occur. The use of unfamiliar terminology without providing adequate definition is a common pitfall. Finding a common language, transparency,

(continued)

Table 4.5 (continued)

Principle of co-creation	Explanation and potential pitfalls
	and clear communication are, therefore, important factors in the co-creation process. Continuous reflection on behaviour and use of language and its effects on the process only becomes possible in a cooperative environment with intense and informal interaction.
Finding attractive value propositions for all	The challenge for co-creation is to find attractive value propositions for all who are involved. This requires design thinking and valuation activities. Valuation can be done through quantitative methods and through the use of words (e.g. sustainable, animal well-being, benefits for the local community)
Fair distribution of costs and benefits	The distribution of costs and benefits should be viewed as fair. This applies to the process and content of co-creation, but also to the outputs, outcomes, and impacts of co-creation. This requires the involvement and strong representation of groups and stakeholders who will be affected or are particularly vulnerable. Stakeholder involvement and 'buy-in', or ownership, is crucial for identifying acceptable trade-offs, for negotiating distributions of costs and benefits and for reaching consensus about the proposed solutions.
Balance between common ground and diversity	Diversity in knowledge, perspectives, and interests supports idea generation but too diverse interests and perspectives might lead to conflict and difficulties in finding common ground. A common pitfall, on one hand, is an incomplete configuration of problem owners and/or knowledge providers. Alternatively, conflicting parties may easily get stuck in a zero-sum game and feel reluctant to find common ground, mutual gains, and joint solutions.
Social learning	Social learning processes at individual and group levels play a central role in co-creation. Examples of social learning are new or adjusted meanings about problems, new technology, social innovations, and societal developments (Sol et al. 2018). A social learning process cannot be divorced from the social context in which it takes place (e.g. see Lave and Wenger 1991). This means that individuals learn by taking part in practice and gradually shift more to the core of the process. However, the practice also influences individuals thoughts and actions, which is common to the structure-agency debate. As Argyris and Schön (1996) have presented, changing values is far more difficult than changing practices since individuals tend to avoid challenging established values.
Iterative learning cycles	Iterative learning cycles usually require longer trajectories of co-creation with sufficient time for reflexive monitoring, evaluation of interventions, and translating those lessons into a new cycle of 'plan-do-evaluate-respond'. A lack of reflexive monitoring (skills) and participation fatigue are common pitfalls. Small wins and good personal relations help to keep people on board for an extended period of time.

(continued)

Table 4.5 (continued)

Principle of co-creation	Explanation and potential pitfalls
Shared ownership	Shared ownership of the process and produced knowledge and outputs determines to what extent participants are willing to cooperate, contribute, share knowledge and follow-up, and implement collective actions or results of the co-creation trajectory. This relates to knowledge management, equal access to and sharing of information and know-how, intellectual property rights (IPRs), and responsibility for maintaining confidentiality (e.g. in case of Chatham House rules), which are all factors that can limit or support a collaborative environment.
Expectation management	Expectation management requires transparency about the process, time-planning, rules of the game (e.g. decision-making procedures or Chatham House rules), different roles, ownership, expected results, and follow-up.
Path dependency	Human behaviour gets shaped to a large extent by routines resulting from choices made in the past and institutional structures. This path dependency is a reason for institutional stability since institutional pressures force organizations to adopt similar practices or structures to gain legitimacy and support (DiMaggio and Powell 1983, 2000). As a result, these institutions can become firmly rooted in taken-for-granted rules, norms, and routines (Seo and Creed 2002). As such, path dependency may limit creativity and out-of-the-box thinking in a co-creation trajectory. On the other hand, the disqualification of 'old' frames of reference—by excluding (too easily) participants who adhere to dogmas, standard practices, or rules of behaviour—could limit equal participation. By emphasizing the historical and contextual systematic character of former rules of interaction it is possible to respect the involved participants and prevent disqualification of their 'old' frames of reference (valuation). This, in turn, might remove the defensive reactions that usually contribute particularly to locking up the existing frame even more firmly. New forms of value make it possible to expand the partnership.
Power of imagination	A tangible and joint vision could serve as a vehicle to identify and create shared and common values during the co-creation trajectory, which is an important requirement to hold actor coalitions together. The ability to form mental pictures or ideas in the minds of participants is perhaps the most powerful tool for co-creation and could be stimulated by methods that support imagination and visualization of a shared future.
Institutional work	Co-creation activities, especially those with transformative impacts, require new partnership, agreements, standards, and activities aimed at adjusting formal and informal institutions.

Open Access This chapter is licensed under the terms of the Creative Commons Attribution 4.0 International License (http://creativecommons.org/licenses/by/4.0/), which permits use, sharing, adaptation, distribution and reproduction in any medium or format, as long as you give appropriate credit to the original author(s) and the source, provide a link to the Creative Commons license and indicate if changes were made.

The images or other third party material in this chapter are included in the chapter's Creative Commons license, unless indicated otherwise in a credit line to the material. If material is not included in the chapter's Creative Commons license and your intended use is not permitted by statutory regulation or exceeds the permitted use, you will need to obtain permission directly from the copyright holder.

Part III
A Research and Innovation Agenda

The core philosophy of the research and innovation agenda presented in this book can be described as developing a powerful combination between practice-driven collaborative action research and theoretically informed scientific research. Collaborative action research means that we take guidance from practice as the primary source of questions, dilemmas, and empirical data regarding Transformative Social-Ecological Innovation, and collaborate with stakeholders in testing insights and strategies, and evaluating their usefulness. Scientific quality will be achieved by placing this co-production of knowledge in a well-founded and innovative theoretical and analytical framework for Transformative Social-Ecological Innovation as presented in Sect. 3.9 and in Sects. 5.1–5.3.

The aim of this research and innovation agenda is to diagnose and advance Transformative Social-Ecological Innovation (TSEI) across sectors and disciplines, and at different levels of governance. The TSEI-framework presented in this book helps to identify intervention points and to formulate systemic and sustainable solutions that can include different views, as well changing and competing needs. Overall, the concept of TSEI encourages public officials, business leaders, and the greater public to think more broadly about how society can rethink cooperation to address humankind's greatest challenges.

The Research and Innovation Agenda presented here will seek to answer the following key research questions, while stressing the importance of engaging in iterative and adaptive cycles of research planning and prioritization with a collaborative network of knowledge partners and partners from practice during implementation of this agenda:

- What governance and policy arrangements are required for realizing a transformation toward a sustainable, healthy and just society?
- In which ways does Transformative Social-Ecological Innovation (TSEI) contribute to a better understanding and realization of the sustainability transition and the societal transformation towards a Natural Social Contract? This primarily includes governance research into the emergence, development, and context of

partnerships between various stakeholders and the extent to which they achieve the intended results.
- Which factors determine the success and/or failure of these TSEIs?
- What is required to enable a just and fair transformation, in particular related to inclusivity, socio-economic inequities, procedural justice, social and environmental justice, legitimacy, potential winners and losers, and potential trade-offs between social, economic, and ecological interests and values?

In the broad and diverse field of sustainability transitions and transformation research (see Köhler et al. 2019 for an overview), the research and innovation agenda presented here, and in the following chapters, can be positioned and characterized as follows:

- Core focus on the Governance of Sustainability Transitions, with specific attention for issues of politics power and justice in transitions, and drawing on the wider field of governance, innovation and transition studies as well as other fields like complexity theory and systems theory.
- Commitment to research that not only describes societal transformation processes, but initiates and catalyses them (Luederitz et al. 2017; Köhler et al. 2019). Various methodological approaches of real-world experimentation and collaborative action research are being used that make the commitment to knowledge co-production operational (see Sects. 4.10, 4.11 and 5.4 in particular).
- Engagement with Transformative Social-Ecological Innovation (TSEI) in-the-making, not only focusing on how to govern the early stages of the process (e.g. transition arenas, niche-experiments), but also later phases of transition (for example, how to achieve acceleration, e.g. see (Gorissen et al. 2018; Sovacool 2016). This primarily includes research into the emergence, development, and context of the partnership between various actors and the extent to which they achieve the intended results. It requires research on what holds the actor coalitions together (e.g. shared beliefs and values, shared discourses, common interests, multiple value creation, and procedural justice), the roles of intermediaries in governing, facilitating, and accelerating transitions, and the role and influence of policy mixes (rather than studying single policy instruments).
- Attention for TSEI outputs, outcomes, and impacts, in particular related to the ethical and normative aspects of a Natural Social Contract, such as social justice, equity, social and environmental stewardship, human security, environmental security, and planetary health. This requires development of appropriate indicators for measuring the social, ecological, and economic dimensions of sustainability transitions.

The activities of this research group are geared towards generating greater insight into and awareness of Transformative Social-Ecological Innovation (TSEI) by means of applied research and education. The resulting knowledge and skills will be used to support and engage with TSEI in-the-making, which in turn will generate new knowledge and skills (i.e. in iterative learning cycles). For this purpose, several analytical instruments for studying Transformative Social-Ecological Innovation will be presented in Chap. 5, including:

- Analytical framework for Transformative Social-Ecological Innovation (Sect. 5.1)
- Power and network analysis (Sect. 5.2)
- A framework for analysing different levels of collective learning (Sect. 5.3)
- A collaborative action research methodology (Sect. 5.4)

Relevant and ongoing research and educational activities will be highlighted in Chaps. 6 and 7, including:

- Chapter 6: Transition to a sustainable and healthy agri-food system
- Chapter 7: Governance of urban sustainability transitions

References

Gorissen, L., Spira, F., Meynaerts, E., Valkering, P., & Frantzeskaki, N. (2018). Moving towards systemic change? Investigating acceleration dynamics of urban sustainability transitions in the Belgian City of Genk. *Journal of Cleaner Production, 173*, 171–185.

Köhler, J., Geels, F. W., Kern, F., Markard, J., Onsongo, E., Wieczorek, A., Alkemade, F., Avelino, F., Bergek, A., Boons, F., & Fünfschilling, L. (2019). An agenda for sustainability transitions research: State of the art and future directions. *Environmental Innovation and Societal Transitions, 31*, 1–32.

Luederitz, C., Schapke, N., Wiek, A., Lang, D. J., Bergmann, M., Bos, J. J., Burch, S., Davies, A., Evans, J., Konig, A., Farrelly, M. A., Forrest, N., Frantzeskaki, N., Gibson, R. B., Kay, B., Loorbach, D., McCormick, K., Parodi, O., Rauschmayer, F., Schneidewind, U., Stauffacher, M., Stelzer, F., Trencher, G., Venjakob, J., Vergragt, P. J., von Wehrden, H., & Westley, F. R. (2017). Learning through evaluation – A tentative evaluative scheme for sustainability transition experiments. *Journal of Cleaner Production, 169*, 61–76.

Sovacool, B. K. (2016). *The history and politics of energy transitions: Comparing contested views and finding common ground (No. 2016/81)*. WIDER Working Paper.

5 Analytical Instruments for Studying TSEI

The findings from literature in the previous chapters have been brought together in a conceptual framework (see Sect. 3.9) and an analytical framework for Transformative Social-Ecological Innovation (Sect. 5.1). It's main purpose is to study the dynamic interplay between actors and institutional structures influencing and inducing institutional change. This chapter furthermore provides a further operationalization of the TSEI analytical framework for analysing shifts in power dynamics (Sect. 5.2), by investigating a series or cluster of closely related action situations and mapping how power dynamics change. An example of TSEI-framework application is provided in Sect. 6.6. Finally, Sect. 5.3 provides a framework for analysing different levels of collective learning, which is considered as one of the key variables for studying the outputs of TSEI. Finally, this chapter highlights some important insights on collaborative action research and related methods (Sect. 5.4).

5.1 Analytical Framework for Transformative Social-Ecological Innovation (TSEI)

The TSEI-framework presented here is based on earlier work by Huntjens et al. (2016) and Huntjens (2019). Predecessors of the TSEI-framework have been used successfully in environmental diplomacy and mediation processes in various parts of the world (Huntjens et al. 2014a, b, c; Yasuda et al. 2017a, b, 2018), as well as for studying transformation processes and institutional change in water management, agriculture, and spatial planning (Islam and Madani 2017; Yasuda et al. 2018, 2020; Huntjens 2019; Huntjens et al. 2020).

Within the TSEI-framework the action situation has been taken as the core object of analysis, and considers the action situation as the interface or 'glue' between two other analytical components: structure/institutions on the one hand, and actor-agency on the other. As such, the framework can be used for institutional and political-economy analyses, with a special focus on the power dynamics at play (Sect. 5.2).

Power dynamics can be studied by looking at a series or clusters of closely related action situations, in which the initiation, format, content, and output of each action situation are analysed.

This framework allows to zoom in on a series or cluster of related action situations (and their context), looking at 'structure' and 'agency', and at the output-outcomes-impact of these action situations (per action situation and per series/cluster). An action situation is a moment where multiple parties (with different interests, perspectives, and preferences) come together and are confronted with a series of potential actions, in which these parties exchange goods and services, try to solve problems, influence each other, learn together, and which results in shared output and outcomes. A series or cluster of closely related action situations is often referred to in literature as an action arena (Ostrom 2009) or transition arena (Loorbach 2010).

For analysing a series or cluster of closely related action situations it is valuable to make use of a learning history or timeline method (Sect. 4.7), because it aims to provide better insight into a series of action situations and the associated learning history. The timeline method is therefore an important part of the methodology of the TSEI-framework. Based on empirical data, for instance based on interviews and timeline method, action situations can be identified for further analysis, in particular those that influenced or were decisive for the process of multi-party collaboration and its results.

The TSEI-framework distinguishes five main components, corresponding to the numbers in Fig. 5.1:

1. TSEI context and situation-specific context
2. Action situation
3. Structure/institutions
4. Actors/agency
5. Outputs, outcomes, and impacts

Each component will be briefly explained below.

Component 1: TSEI Context and Action Situation Context
Understanding the circumstances that influence the nature of the Transformative Social-Ecological Innovation and those that affect a decisive moment in the cooperation process (the action situation) is an important first step in the analysis. Examples of contextual factors include the nature and extent of the societal change, the history of cooperation between the parties involved in past action situations (or the lack thereof), and the key biophysical, material, and socio-economic features of the area in question, such as a rural or urban district, a province or ecoregion.

Component 2: The Action Situation
An action situation is a situation in which two or more individuals are confronted with a series of potential actions that will result in shared outputs and outcomes

5.1 Analytical Framework for Transformative Social-Ecological Innovation (TSEI)

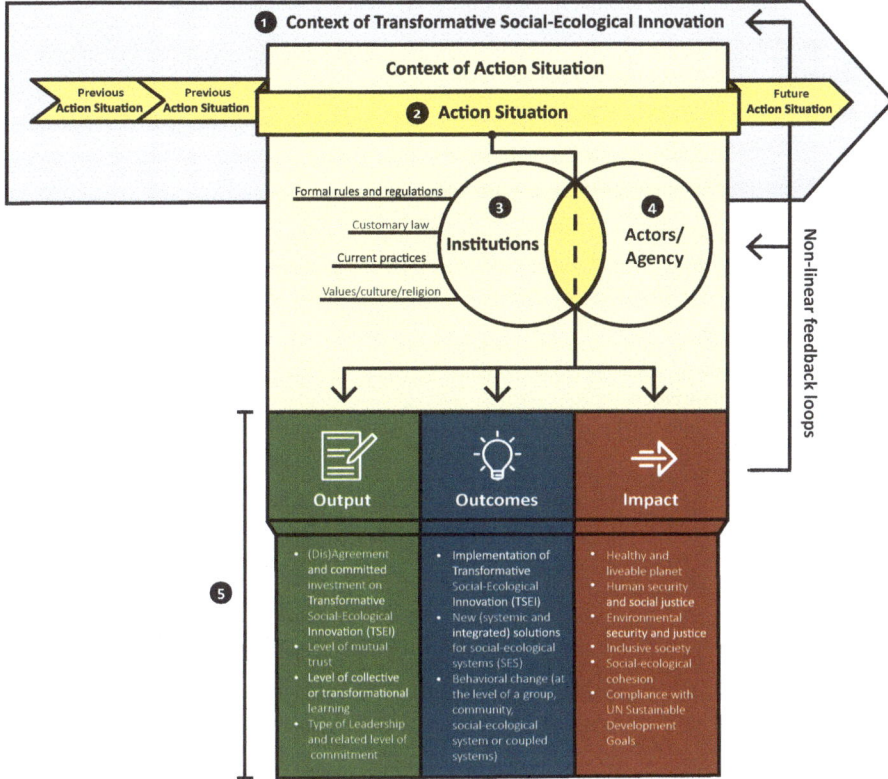

Fig. 5.1 Analytical framework for Transformative Social-Ecological Innovation (TSEI). Numbers in this figure correspond to description of specific components below

(Ostrom 1999, volume 42; 2005, p. 13). An action situation is referred to as 'the social space where participants with diverse preferences interact, exchange goods and services, solve problems, dominate one another or fight' (Ostrom (2005, 32). Researchers using the TSEI-framework may select an action situation by examining whether the situation in question is or was decisive for the process of cooperation and/or its outcome. This can range from multi-stakeholder dialogues to meetings within a negotiation or decision-making process, often as part of a series or cluster of closely related meetings or negotiations. It is often necessary to study several different action situations, as well as their relationship to each other, in order to gain a better understanding of the TSEI. These series of clusters are also referred to as action arenas or transition arenas. The selected action situations are then analysed, focusing in particular on subcomponents such as initiation, process, format, and content of the action situation. Detailed questions regarding these subcomponents are listed in Table 5.1, based on Huntjens et al. (2016, 2017) and Huntjens (2019).

Table 5.1 Subcomponents and questions for the analysis of the action situation

Subcomponent	Question
Initiation	• What triggered the meeting? • What was the objective? • Who organized it? Who was invited, who was not, and why? • How was support mobilized?
Process/ Format	• Who was present and who cancelled? • Were there any specific reasons for participating or cancelling? • Which venue was used and how was the meeting structured (agenda)? • Who acted as a facilitator? How was inter-participant exchange facilitated? • Which discussion format was used, e.g. round-table discussion, a workshop, or a more advanced participation method? • Who spoke and who took minutes? • Was there any expectation management and was the decision-making process transparent? • Which decision-making protocol was used, e.g. majority vote, consensus, consent? • Which negotiation strategies were used, e.g. accepting the first offer, compromising (splitting the difference), competition (zero-sum game), or problem solving (mutual profit)?
Content	• Which issues and topics were addressed during the action situation? Which were excluded or avoided? • What information was made available to participants in advance? Was it relevant? Was there enough time to take in this information? • Which uncertainties were identified and/or addressed in the action situation? • Did participants allow their knowledge and information to be challenged by other participants and did they present their own mental models, insofar as they were aware of them? • Was information presented in an authoritative way or a facilitating way, encouraging other participants to reflect? • Did new information emerge during the action situation, and how did this affect the negotiations or dialogue?
Output	• Agreements and related level of commitment, mutual trust, level of collective or transformational learning. For more information, see component 5.

An example highlighted in this book (Sect. 6.6) is a series of closely related action situations, including the final adoption of the Ambition document on the Innovation Agenda for Sustainable Agriculture by the Provincial Council of Zuid-Holland on 29 June 2016. Although the adoption did indeed conclude a decision-making process with regard to ambitions and the agenda, it mainly constituted an important step within a longer-term process of change towards a strong, sustainable and future-proof agriculture and food chain in the Province of Zuid-Holland. With the adoption of this document, the Province made seven million euros in co-financing available, in addition to seven million euros in European subsidies from the Rural Development Programme (In Dutch: Provinciaal Ontwikkelingsprogramma), adding up to a total of 14 million euros in available funds. Entrepreneurs can use this funding to implement innovations in experimental projects to drive sustainable agriculture. In addition, 350,000 € of co-funding were set aside for the Knowledge and Development programme, an initiative by various educational institutions and universities to collect and share knowledge. To facilitate the transition approach and network building, approximately 650,000 € have been made available for a period of 4 years.

Component 3: Institutions
The concept of 'institutions' has several different interpretations in literature. This book follows the definition proposed by Calhoun (2002, p. 33): 'Institutions are deeply rooted patterns of social practices or norms that play an important role in how society is organised'. Institutions can pertain to various areas of social activity, such as family life, associations, and politics. Generally speaking, institutions result from a process of institutionalization, in which preferences are gradually strengthened until they are fixed and familiar. This process is usually accompanied by conflicts and the exercise of social power (Parker et al. 2003). We distinguish between formal and informal institutions:

- **Formal institutions** are those that structure the practices of actors and which are adopted through a formalized process. They include the constitution, laws, and legislation adopted by society, organizations, and policy.
- **Informal institutions** are those that structure the practices of actors and which are embedded in organizations or groups without a formalized process. They include customary law, existing practices, norms, and culture.

Component 4: Actors/Agency
Agency refers to an actor's ability to exert influence (Ali-Khan and Mulvihill 2008; Newman and Dale 2005). The first step in analysing this component consists of identifying key stakeholders and actors, with the former referring to all persons, groups, and organizations with an interest in the societal change in question, either because they are affected or because they can influence its outcome. This may include individual citizens and businesses, interest groups, government agencies, and experts. It is important to map the interests, incentives, and access to financial, personal, or institutional resources of all stakeholders who participate actively in the action situation. On top of that, existing coalitions and partnerships need to be taken into account in the analysis, since they can influence the power dynamics. In order to better understand cooperation and decision-making, it will often be necessary to identify the preferred or dominant negotiation and influence strategies of each actor, as this information, when bundled, will provide greater insight into the role and influence of each individual actor.

Cooperation requires potent leadership and management (Leach and Pelkey 2001; Huntjens 2011), which is why it is important to understand the leadership styles in play. Leadership has an important role in building trust, substantive management, conflict management, connecting parties, initiating cooperation, collecting and generating knowledge, and mobilizing broad support for change (Folke et al. 2005).

Component 5: Outputs, Outcomes, and Impacts
An action situation can result in outputs, outcomes, and impacts, three distinct concepts. The difference between these three is defined as follows (Huntjens 2019):

- **Output:** the product resulting from one action situation or output of series or clusters of closely related action situations. Examples of output include a cooperation treaty, other types of agreement, committed investment, a plan, strategy, legislative proposal, financial regulations, or instruments to promote sustainability. Also the level of collective or transformational learning (see Sect. 5.3), mutual trust, type of leadership, and related level of commitment are considered as outputs.
- **Outcome/result:** this is the direct effect of the output. It is measurable and time-limited, though determining the full effect can take an extended period of time. Examples of outcomes include behavioural change, new knowledge, and solutions resulting from co-creation and social learning. A new revenue scheme for sustainable business or a circular business model could be outcomes of (new) financial regulations and instruments that promote sustainability.
- **Impacts:** these are the long-term or indirect effects of the outcomes/results.

Impacts are often difficult to quantify because they may or may not happen. Impact is what we hope for, whereas results are what we work for. To illustrate the difference between results and impact: In sustainable business practices aimed at nature-inclusive agriculture, farmers work to make a living (result), and with biodiversity measures (also result) they hope for the restoration of biodiversity (impact). When creating green spaces or water collection facilities in the city (result), residents and other parties hope for improved air quality and better protection against flooding (impact).

It is also important to consider unintended side effects, as it is possible for policy to achieve its intended goals while also leading to a large number of adverse side effects (Biermann et al. 2007). The introduction of phosphate rights in the Netherlands, for instance, had unintended negative side effects, such as the irresponsible increase in milk production per cow and the significant growth of dairy farms without using extra land.

5.2 Power and Network Analysis

Because social innovation is a process that involves several groups, each of which have their own norms, values and interests, issues of distribution and power are inevitable (Meadowcroft 2009; Cattacin and Zimmer 2016; Karré 2018). However, the balance of power and the interests that play a role in social innovation often remain underexposed, while the question of how to deal with competing interests and values—and how to use this competition to prompt co-creation—plays a crucial role in the success or failure of cooperative efforts. As such, it is important to pay

5.2 Power and Network Analysis

attention to the role of power and what influence it has on the decision-making process.

A power analysis or network analysis (see, for example, Wielinga and Robijn 2018) is therefore an indispensable instrument in order to better understand and facilitate social innovation and effective cooperation. A power or network analysis is preferably carried out together with the actors involved, with participatory analysis thus contributing to mutual understanding and the process of social learning and co-creation.

According to the philosopher Nagel (1975), power is a causal link between a party's wishes for a result and the result itself. It is distinct from sources of power, from which power can be drawn. Power comes in all sorts of shapes and sizes, including potential power, latent power, implicit power, and manifest power. Most often, power is exercised implicitly: the most powerful player does not make a threat, but others still take the threat they may pose into account. Other types of power include process power, structural power, and coercive power, the latter of which leans on persuasion and has the potential to harm others. Powers and influences are two sides of the same coin: influencing the behaviour of others.

The work of Partzsch (2017) informed us on three ideal type concepts: 'power with' (learning and cooperation), 'power to' (resistance and empowerment), and 'power over' (coercion and manipulation). Furthermore, the multi-level power framework offered by Arts and Van Tatenhove (2004, based on Clegg, 1989) distinguishes between relational, dispositional, and structural power. Avelino and Wittmayer (2016) argue 'that besides such a vertical typology of power, as offered by Arts and Van Tatenhove (2004), we also need a horizontal understanding which allows to analyse who exercises relational power (in a specific action situation), and also, how the dispositional power embodied in actor configurations is configured across different actors'.

The following explains how the TSEI-framework (see Sect. 5.1) can be used to analyse the role of power within transformative social-ecological innovations (TSEI). To this end, a distinction is made between three forms of power that can play a part in the process of TSEI at different levels: relational, dispositional, and structural power (based on Arts and Van Tatenhove 2004):

- **Relational power**: an actor's capacity to achieve its goals in interaction with other actors. Power can only be expressed in social relationships, and at this level, actors and their motivations, resources, interaction, and outputs are central. In a process of social innovation, relational power can be expressed in an actor's capacity to put certain problems on the agenda and framing them, or their ability to mobilize resources to achieve the desired change. Actors can have various motives for innovation, such as changing circumstances and shifting perceptions, shock events, and problems with existing policies. Professors Bas Arts and Jan van Tatenhove (Arts and Van Tatenhove 2004) do, however, qualify this by adding that human behaviour is shaped to a large extent by routines, path dependence resulting from choices made in the past and institutional structures (Arts and Van Tatenhove 2004).
- **Dispositional power**: an actor's capacity to act. Actors are positioned in organizational structures which give them a certain degree of access to resources, but informal norms and formal rules can also affect an actor's freedom of action and behaviour. Dispositional power is expressed by seemingly 'fixed' organizations or institutions but is certainly not static. After all, actors form institutions just as much as they are affected by them. According to renowned sociologist Anthony Giddens (1984) and political scientist Alexander Wendt (1987), actors have preferences that they cannot realize without collective action. Based on these preferences, they form and reform certain social structures over time, affected partially by unintended consequences (see Grin 2010). Once these social structures are in place, they begin to give direction to actors themselves and their preferences. Actors are capable of changing the organizations they work in, but due to their duration these processes transcend daily politics (Arts and Van Tatenhove 2004).
- **Structural power**: refers to the nature of signification, legitimation, and distribution of power in a society and constitutes how macro-social structures, such as discourses and institutions, influence actors. Structural power stimulates certain outcomes of interactions or processes while hampering alternatives that conflict with prevailing discourses/institutions. Structures are not actors, of course, but they are reflected in the behaviour of actors, which is why structural power can also be identified when researching the motivations of individual actors. Structural power is subject to change, but these processes are slow and often last longer than a human lifetime (Arts and Van Tatenhove 2004).

Grin (2011) has translated these three forms of power into the *multi-level perspective* of transition science, describing relational power, for instance, as an actor's capacity to use the regime to their advantage. Dispositional power is represented in the regime and its formal rules, access to resources, configurations of actors and dominant norms or ideas. Structural power is expressed at the landscape level and influences what is desirable and legitimate.

Power is therefore an inherently dynamic and layered concept (Avelino and Rotmans 2009, p. 559; Grin 2016, p. 112). These three distinct forms of power make up a vertical typology of power, in which the different kinds of power correspond to different degrees of aggregation (actors, structures, systems). Avelino

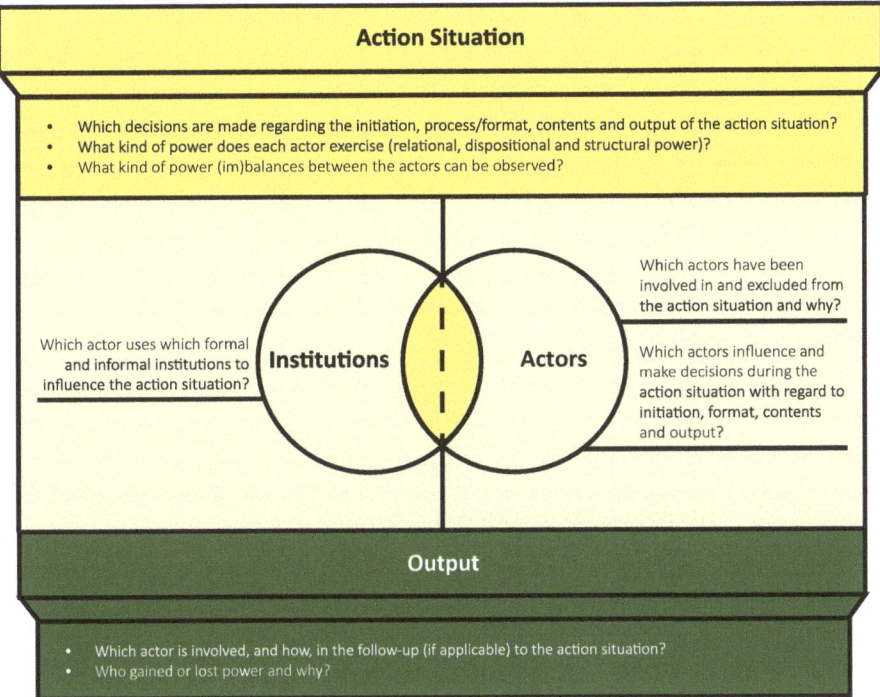

Fig. 5.2 Using the TSEI-framework for power analysis

and Wittmayer (2016) argue 'that in addition to a vertical typology of power, we also need a horizontal approach, distinguishing between three types of power relations between actors: (1) A has power over B, (2) A has more/less power than B to do x, and (3) A and B have different kinds of power'. In essence, Avelino and Wittmayer (2016) ask 'how different actors exercise different kinds of power at different times in different roles' (cf. Avelino and Wittmayer 2016). Both the vertical and horizontal typology of power presented here have been included in the power analysis within the TSEI-framework (see Fig. 5.2). As such, the TSEI analytical framework can also be applied to analyse shifts in power dynamics, by investigating a series or cluster of closely related action situations and mapping how power dynamics change.

5.3 Framework for Analysing Different Levels of Collective Learning[1]

In order to distinguish different learning processes and how to classify them according to the triple loop concept it is useful to start with some definitions (based on Hargrove 2002):

---

[1]This section is based on Huntjens et al. (2011a).

Table 5.2 Collective learning framework (Huntjens 2011; Huntjens et al. 2011a)

Type	Indicators
Single loop learning	1. Small changes are made to specific practices or behaviours, based on what has or has not worked in the past. Things are done better without necessarily examining or challenging underlying beliefs and assumptions (Kahane 2004). Goals, values, plans, and rules are operationalized rather than questioned (Argyris and Schön 1974) 2. Goals, values, frameworks and, to a significant extent, strategies are taken for granted. The emphasis is on techniques and making techniques more efficient (Usher and Bryant 1989, 87).
Double loop learning	1. Modifications (as the result of learning) are occurring, or have occurred, in personnel, programs, and legal and organizational structures that incorporate new information (including policy feedback) and causal understandings that yield more intellectually perceptive processes, a wider range of capabilities, and more effective policy (Brown 2000, 3). 2. Actor networks are changed by including new stakeholders, supporting reflection on assumptions, and showing new possibilities. The social network of stakeholders is seen as the basis for learning and dealing with change (Folke et al. 2005; Geels et al. 2004). 3. Uncertainties are identified as a first step to find solutions (Brugnach et al. 2008), and then taken into account in policymaking (Huntjens et al. 2010).
Triple loop learning	1. Horizons of possibility are expanded (Hargrove 2002, 118). 2. A paradigm shift takes place that alters our way of thinking and behaviour (Hargrove 2002, 119; Pahl-Wostl 2009). 3. A major structural change takes place in the regulatory framework.

- **Single loop learning (SLL)** is a refinement of established actions to improve performance without changing guiding assumptions or taking alternative actions into account.
- **Double loop learning (DLL)** is a change in frame of reference and guiding assumptions.
- **Triple loop learning (TLL)** is a transformation of context to change factors that determine the frame of reference. It refers to transitions of the entire regime in which values and norms are shaped and stabilized by structural context.

The concept of multi-loop learning has been further operationalized into an analytical framework, summarized in Table 5.2 (Huntjens 2011; Huntjens et al. 2011a).

5.4 Collaborative Action Research[2]

For research into social phenomena there is increasing interest in 'action research' in various forms. In this process the researcher enters a real-world situation and aims both to improve it and to acquire knowledge (Checkland and Holwell 1998). Since

---
[2]This section is based on Huntjens et al. (2011b).

5.4 Collaborative Action Research

the 1990s it became more and more difficult to identify the main thrust of action research, since there have been a number of different interpretations of the term action research, but also a variety of different terms, such as action learning, action research, action inquiry, participatory action research, and collaborative action research (Eden and Huxham 1996). All of them share the aim of building 'theories within the practice context itself and test them through intervention experiments' (Argyris and Schön 1978; Argyris 1985).

The need for practical, useful research that informs management practice is well established. For a number of reasons, action research is well suited to provide actionable knowledge (Coghlan and Brannick 2002). Action research provides relevant knowledge due to the involvement of practitioners and because the research is carried out in the relevant context itself. Due to the involvement of practitioners, rich data can be gathered relatively easily. It provides rich data due to the involvement of practitioners. Because data are gathered in context, the research results are valid in that context. The involvement of practitioners enhances the development of actionable knowledge, while scientific researchers in action research tend to guard the development of theoretical knowledge. Action research projects often use both qualitative and quantitative methods, and can provide both theoretical and practical insights (Reason and Bradbury 2005).

Action research aims to contribute both to the practical concerns of people in an immediate problematic situation and to further the goals of social science simultaneously (Gilmore et al. 1986). In other words, there is a dual commitment in action research to study a system and concurrently to collaborate with members of the system in changing it in what is together regarded as a desirable direction. The twofold ambition of developing practically relevant and scientifically sound knowledge requires the active collaboration of researcher and client, and thus it stresses the importance of co-learning as a primary aspect of the research process (Gilmore et al. 1986). Action research involves utilizing a systematic cyclical method of planning, taking action, observing, evaluating (including self-evaluation), and critical reflecting prior to planning the next cycle (O'Brien 2001). Of course, not all problems and research topics require the same standard approach. Each action research programme requires tailor made arrangements, which take—amongst others—into account situational conditions regarding the content of the issues, relationships, and commitments.

The principle of actively involving stakeholders in our research on Transformative Social-Ecological Innovation is important for several reasons. The first reason is that stakeholder involvement and 'buy-in', or ownership, is crucial for identifying acceptable trade-offs, for negotiating distributions of costs and benefits and for reaching consensus about the research findings and recommendations (Ashby 2003). During processes of Transformative Social-Ecological Innovation, the understanding needed for consensus and compliance requires new knowledge to be generated by research in order to achieve stakeholder 'buy-in' and often needs to include expertise drawn from other stakeholder groups (Irwin 1995). This form of ownership often needs to be established across a range of institutions and levels of decision-making (Martin and Sutherland 2003).

A second reason for involving stakeholders in research is that their involvement is a key to coping with the complexities and uncertainties related to Transformative Social-Ecological Innovation, by bringing in a wider range of perspectives on needs, impacts, and options, and having them deliberated openly. At the same time, by engaging with complex governance systems, researchers are better able to understand their dynamics.

The issue of great complexity and uncertainty poses important challenges to governments, particularly in finding their most appropriate role in the governance of sustainability transitions. They try to find answers on questions like: which instruments can we use, which policy options are available, how do we have to organize governance processes and which legal room for manoeuvre do we have? Instead of studying these considerations themselves, collaborative action research can be an approach to help officials by finding the right answers.

A third reason is to use collaborative action research in the emerging field of governance of sustainability transitions is that this field is still in its infancy. Many stakeholders are still thinking about what they have to do and how they have to do this. Hence, there is not so much opportunity for reconstructive research, for in-depth surveys or multiple case-study research when we want to know more about the Transformative Social-Ecological Innovation. We have to focus our research on practices which are emerging.

Fourth, because the theory of a Natural Social Contract and related Transformative Social-Ecological Innovation is under construction, it is very helpful to organize short, iterative cycles of observation, analysis, and adjustment. Action research is highly useful to combine initial theory testing and theory development. It provides in recurring learning cycles in which empirical fieldwork and theoretical reflection follow each other.

It is not the case, however, that intensive, time-consuming participation processes must be organized for each and every problem. Within complex transition challenges, such as the sustainability transition, stakeholder participation, and collaborative action research are needed when:

1. different stakeholders depend on each other to achieve their goals.
2. there is no agreement about the problems at hand or the solutions to these problems.
3. information is incomplete or disputed, with the necessary knowledge and experience being distributed among different parties.
4. the issues at hand are sufficiently important for stakeholders to invest the necessary time (and therefore money) in solving them.

Huntjens (2011) observed that parties involved in complex social problems in practice make only limited use of the broad range of methods in which diverse stakeholders can learn from each other or utilize each other's knowledge and experience. Choosing for an appropriate method for social learning and stakeholder participation can make the difference between confrontation or cooperation between parties. In practice, an enormous number of workshops, inspiration workshops, round-table discussions, and brainstorming sessions are organized, but it is often questionable to what extent the right method or approach has been chosen in order to achieve social learning, effective cooperation, and mutual trust. For each phase and aspect of the cooperation process, many different methods are available to bring parties together, to facilitate social learning, and to realize collective action. In addition to the chosen method, the facilitation style and required competencies are of great importance.

Broad stakeholder participation generally provides a diverse and more nuanced set of measures needed to address complex issues, which also promotes a social system's capacity to learn (Huntjens 2011). However, social learning does not happen by itself and requires effort on the part of all those involved, adequate facilitation and reliable information on issues being discussed. The most suitable method has to be chosen on the basis of the phase in the planning or implementation process, the composition of the group, the context, the objective, the ambitions, and the desired outcome of a meeting.

There are dozens of proven methods to stimulate and facilitate interaction, participation, social learning, and co-creation. Below illustrations are only a small selection of the available methods:

- **Vision development**: a method for reformulating substantive objectives when common ground is lacking and the planning is fragmented (Hajer and Poorter 2005).
- **Role-playing**: gamified simulations can be used to experiment with real-life processes in a somewhat controlled environment (Cook and Campbell 1979; Vissers et al. 1995), and involves people playing various roles in order to imitate the social system. By way of illustration, role-playing helped De Stichtse Rijnlanden water board develop the area plan for the Kromme Rijn (Change Magazine 2009). More space was needed to collect water and ditches had to be widened, which would force the horticultural and agricultural sector to surrender land. When the board proposed its plan, it met with considerable dissent. In a role-playing game, farmers and citizens were asked to take a seat in the water board's boardroom to follow the same decision-making process, which led to a tumultuous meeting that finished with the participants coming up with the same plan as the water board. In the end, the role-playing process increased mutual understanding and trust (ibid.).
- **Group model building**: this method can help identify interdependencies and define a common problem and solutions. All parties sit around the same table and are given an equal opportunity to explain why they believe the policy in question to be successful or unsuccessful, writing it on a large sheet of paper charting the relationships between various factors. This process allows for everyone to be heard, including different ministries, municipalities, provinces, businesses and environmental organizations, as well as individual citizens, fishermen, and farmers. This method therefore creates an understanding of each other's interests while broadening the horizons of all parties involved and giving the entire process

more depth. There is often no such thing as a simple solution (Huntjens et al. 2014a, b, c).
- **Backcasting**: this is a commonly used method in spatial planning that involves imagining a successful future outcome, after which the participants ask themselves what must be done today to achieve this situation (Quist and Vergragt 2006).
- **Reflexive monitoring**: this method involves mapping learning processes and helping project participants reflect in order to help strengthen system innovation projects (Van Mierlo et al. 2010).
- **Learning history/timeline method**: different stakeholders will come to different evaluations of the same project or programme. Exchanging and discussing these evaluations contributes to deeper learning and developing a common perspective on innovation (Willems and Roelofs 2009).
- **Dynamic learning agenda**: this method is used to formulate, record, and keep track of long-term challenges and concrete action perspectives (Regeer et al. 2009; Van Mierlo et al. 2010).

Open Access This chapter is licensed under the terms of the Creative Commons Attribution 4.0 International License (http://creativecommons.org/licenses/by/4.0/), which permits use, sharing, adaptation, distribution and reproduction in any medium or format, as long as you give appropriate credit to the original author(s) and the source, provide a link to the Creative Commons license and indicate if changes were made.

The images or other third party material in this chapter are included in the chapter's Creative Commons license, unless indicated otherwise in a credit line to the material. If material is not included in the chapter's Creative Commons license and your intended use is not permitted by statutory regulation or exceeds the permitted use, you will need to obtain permission directly from the copyright holder.

Transition to a Sustainable and Healthy Agri-Food System   6

6.1 Challenges and Developments

Agriculture, horticulture, aquaculture, and fisheries are essential for our food production and therefore indispensable in our society. They are an integral part of our economies and cultures. By 2050, the world will have a population of about nine billion, with rapidly changing nutritional needs. With the vast majority of consumers usually opting to pay the lowest price, it prompts the food industry to adopt highly efficient, low-lost production methods. As a consequence, there is little incentive for actors in the food chain to invest in sustainability measures and translate those into cost price. This economic logic leads to a vicious circle.

Current food consumption and production patterns contribute strongly to a number of urgent sustainability challenges in the areas of health and well-being of humans, animals, and the planet. The global food system is under great pressure, due in part to the growing world population and climate change, but also because of how we currently produce and consume food. The Agri & Food sector has traditionally focused on production and efficiency, producing as much food per square metre as possible at the lowest possible cost and with a limited view of value creation. The predominant focus on productivity, the free market, and profit maximization has shifted social and ecological values and costs to the background. Profit is narrowly defined in monetary terms by externalizing ecological and social costs, which means these 'hidden costs' are usually not reflected in the price of food. A recent estimate puts the 'hidden costs' of global food and land-use systems at $12 trillion, which is 20% more than its market value of $10 trillion (Pharo et al. 2019).

These 'hidden costs' can be grouped into two broad categories:

- **Planetary/ecological costs**: The global food system contributes directly to exceedance of four planetary boundaries: climate change, loss of biodiversity, unsustainable land and water use, and excess nitrogen and phosphorus production (Steffen et al. 2015; IPBES 2019; Willett et al. 2019). Our current global food production and consumption is the single largest greenhouse-gas-emitting sector in the world (IPCC 2019). Moreover, the global food system is by far the largest cause of biodiversity loss, terrestrial ecosystem destruction, freshwater consumption, and waterway pollution due to overuse of nitrogen and phosphorus (IPBES 2019; Rockström et al. 2020).

- **Societal costs**: Globally, more than 820 million people have inadequate access to food, 2 billion people suffer from micronutrient deficiencies (i.e. living on a diet lacking in iron, vitamins, or other micronutrients such as iodine), 600 million people become ill every year from consuming contaminated food causing 420,000 people to die each year, including 125,000 children under the age of 5 year (Havelaar et al. 2015). Another disturbing fact is that the absolute number of undernourished people continues to increase for several years in a row, which makes it very unlikely that the Sustainable Development Goal of Zero Hunger will be accomplished in 2030. The trends of overweight and obesity give additional reason for concern, as they continue to rise in all regions. The most recent data show that obesity is contributing to four million deaths globally and is increasing the risk of morbidity for people in all age groups (FAO, IFAD, UNICEF, WFP, and WHO 2019). Unhealthy dietary patterns lead to an increase in type II diabetes, cardiovascular disease, and some cancers. Six of the eleven disease and mortality risk factors are related to food (Glopan 2016; EAT Lancet 2018).

More and more people are realizing that the current food system will have to be made more sustainable and healthier, but the agricultural sector is lacking in its ability for self-regulation and capacity for governance (Keulartz and Pekelharing 2019). The majority of farmers are stuck in the existing system, and in order to keep their heads above water, many farmers have had to resort to accumulating large debts. They are chained to the banks who fund their businesses and expect farmers to increase their scale and yield. Because of the major pressure put on the system, there is a great deal of disagreement about where to go next within the sector, as witnessed by the massive farmer protests in the Netherlands and other EU countries in 2019, in response to new policies for reducing nitrogen emissions.

Above issues explain why the license to produce for the food sector is under great pressure, while at the same time there is increasing demand for agricultural products. Critical reviews of the agricultural system (e.g. Janssen and Erisman 2016; NewForesight and Commonland 2017; Godfray et al. 2010; SAPEA 2020) warrant a radical transition to a sustainable and healthy agri-food system. The evidence reviewed in the report 'A Sustainable Food System for the European Union' confirms the view that radical system-wide change is required, with 'business as usual' no longer a viable option (SAPEA 2020). There is an urgent need for a new Common Agricultural Policy that goes beyond economic efficiency and maximum production, and also adopts a broader understanding of prosperity. We need an approach that accounts for the climate, nature, animal well-being, public health, and the environment from the outset, rather than requiring retrospective repair legislation (Keulartz and Pekelharing 2019).

6.1 Challenges and Developments

The Netherlands Ministry of Agriculture, Nature and Food Quality has recently developed a new policy for circular agriculture (LNV 2018), with Minister Schouten explaining: 'Farmers want directions, so that's what I'm giving them', in an interview with Trouw newspaper (8 September 2018). However, the question remains whether this new policy will usher in a transition or whether it will simply make minor adjustments and efficiency improvements to the existing system. One of the fundamental questions is to what extent a policy targeted at circular agriculture is compatible with the promotion of free trade in WTO and GATT negotiations and other fora, in particular promoted by the European Union, the USA, and Japan, while at the same time practicing protectionism and subsidies for the domestic agricultural sector (Otero et al. 2013). The European Union's Common Agricultural Policy (CAP) currently accounts for 37.8% of the total EU budget (Rockström et al. 2020). Rockström et al. (2020) suggest to shift these types of subsidies to reward the production of public goods (such as carbon capture, habitat creation, and improved water quality) for securing the global commons while supporting farming communities.

Within this context, several important trends that may either limit, support, or influence, a transition to a sustainable, healthy and just food system can be identified (in no particular order and non-exhaustive):

- **More sustainable purchasing behaviour among consumers**: consumer awareness on sustainability is on the rise, although consumer awareness does not translate directly to sustainable purchasing behaviour (Logatscheva 2016). Nevertheless, studies from various continents show that more and more consumers take sustainability into account when buying items (Joshi and Rahman 2017; Heo and Muralidharan 2019; Logatscheva 2019). The Sustainable Food Monitor in the Netherlands notes that consumer spending on sustainable food was 4.9 billion Euro in 2018: an increase of 7% compared to 2017, while the share of sustainable food in total food expenditure was 11% in 2018 (Logatscheva 2019). If this trend continues and consumers consistently start buying more sustainable products and adapting their consumption patterns, demand for sustainable agricultural products will increase at the expense of conventional products. For instance, this trend has resulted in larger market shares for organic and biodynamic supermarket chains such as Ekoplaza and Odin in the Netherlands (Distrifood Dynamics 2019). At the same time, this development calls for better methods for measuring sustainability in food chains to ensure transparency for both producers and consumers (see Sect. 6.4).
- **Climate change:** The boomerang-effect being observed is that the global food system is the single largest greenhouse-gas-emitting sector in the world (IPCC 2019), while climate change will have significant negative impacts on food security (Dawson et al. 2016; IPCC 2019). 'Climate change will affect all four dimensions of food security: food availability, food accessibility, food utilization and food systems stability' (cf. FAO 2016). It will have an impact on human health, livelihood assets, food production and distribution channels, as well as changing purchasing power and market flows (ibid.). People are becoming

increasingly aware of climate change, and the need to do something about it. Awareness of the causes and impacts of climate change has made society call for a more sustainable and climate-resilient agri-food system.

- **Loss of biodiversity**: biodiversity is essential for pollination, seed dispersal, climate regulation, pest control, biomass production, nutrient recycling, water recycling, soil formation, and soil retention. As soon as the biodiversity within agricultural systems as well as in ecosystems declines, their resilience diminishes with it. Awareness of the correlation between biodiversity and food production, and related vulnerabilities and opportunities, will help set the transition in motion, particularly with a view to preserving and restoring biodiversity in and around agricultural systems. Nature-inclusive agriculture and regenerative agriculture are important developments in this context (Sect. 6.2).
- **Globalization**: The global food system has become industrialized and food chains have become longer, more complex, and more international. It runs in parallel with the growing importance, influence, and vested interests of large industrial players, such as seed and feed suppliers, food processing and packaging industry, transport and logistics, and chains of supermarkets and restaurants. As a part of this development, a large part of the markets has been taken over by a small number of companies (WRR 2014). The power of the few is only increasing as a result of globalization and a capitalist economic logic. Consequently, for consumers it is virtually impossible to have an influence on the range of products on offer in supermarkets, while farmers are forced to comply with the demands of a very small group of purchasing organizations. Farmers are increasingly locked-in the system due to increased specialization and investments in one product, one production method, and one market.

  Within this context, the Corona-crisis has revealed a number of vulnerabilities to the current global model of unlimited circulation of goods and people. In the short term, the protectionist measures and closure of borders have caused a significant disruption of global agro-food chains, for example, causing a shortage of foreign seasonal workers, destruction of fresh produce due to market failures, or confronting export-dependent farmers with lower prices through their contracts with purchasing companies, wondering whether those contracts are not unilaterally passing the risks on to them. On the medium term, the economic crisis following the Corona-crisis will have major consequences for the global food system for the years to come. It will also provide a window of opportunity for reform and interventions that were first conceived impossible.
- **Local food systems**: In recent decades, the distance between farmers and citizens has increased, due in part to urbanization, the anonymity of the supply chain, and the increase in scale, which means that consumers hardly know anything about where their food comes from and that farmers no longer feel valued. As a kind of correction mechanism for the economic climate of liberalization, privatization, and globalization, often in combination with environmental concerns, a growing number of food producers and consumers are opting for shorter food chains and local food systems. Examples include food networks, short food chain platforms, citizens' farms, urban farming, and community-based agriculture, in which

farmers and consumers work together to produce their own food without relying on wholesalers and supermarkets. It enables farmers to receive a fair price for their produce while at the same time restoring the farmer–citizen relationship as well as the relation between humans and nature (see Sect. 6.3).

- **Human health**: Health is becoming an increasingly important topic in society now that global obesity rates have doubled since the 1980s and other diseases are becoming considerably more prevalent. The number of people with type-2-diabetes, for instance, has quadrupled over the past 30 years. These developments can mainly be blamed on poor, unhealthy food and diets. People need more information about the origins of food, its nutritional value, its contents, nutrients, and minerals. However, various scandals within the Dutch food industry, such as the fipronil crisis, have left consumers losing confidence in reliable information. In addition, society is becoming more interested in personalized food, food apps, and food blogs, with food gurus such as Jamie Oliver, Nigella Lawson, and Yotam Ottolenghi having tremendous impact on consumers.
- **Social justice and poverty**: The current food system is not considered fair enough from different perspectives (Alkon and Agyeman 2011). First, equitable access to healthy food is a critical challenge, in particular for lower-income groups (Power 1999; Wertheim-Heck et al. 2019), with more than 820 million people worldwide having inadequate access to food (Havelaar et al. 2015). Second, there are concerns, both within and outside Europe, about wages and working conditions in food chains. 'A condition for the proper functioning of the agricultural system is that the farmers and other agricultural workers who are directly responsible for food production can make a decent living' (cf. NewForesight and Commonland 2017). However, more than three-quarters of smallholder farmers in developing countries are caught up in a poverty trap, which is a trap of low wealth that is virtually impossible to exit because of exclusion from financial markets and an inability to reduce consumption and engage in even a modest savings strategy (Tittonell and Giller 2012). Third, gender inequality and other social inequities are important topics that are often neglected in the analysis and governance of the food system (Schipanski et al. 2016). In general, more attention is required for strategies and conditions that enable a just and fair transformation towards sustainability, in particular related to inclusivity, socio-economic inequities, procedural justice, social and environmental justice, legitimacy, potential winners and losers, and potential trade-offs between social, economic, and ecological interests and values. For instance, evidence suggests there will be winners and losers in reforms to the EU Common Agricultural Policy (CAP) budget and related subsidies (SAPEA 2020; Boulanger and Philippidis 2015; Larrubia Vargas 2017).
- **Animal health and welfare**: All aspects of animal diseases and well-being of food producing-animals during breeding, rearing, transportation, and slaughter have become increasingly important in recent years. The One Health concept is a worldwide strategy for expanding interdisciplinary collaborations and communications in all aspects of health care for humans, animals, and the environment (Kaplan et al. 2009). The main premise is that if you keep animals

under healthy conditions, it is also beneficial for people. Visseren-Hamakers (2020) argues that animal health, welfare, and rights should be added as an 18th Sustainable Development Goal (SDG), considering the neglect of animal considerations in discussions on sustainable development.
- **Security and conflict**: global food system challenges must be viewed from an international perspective and the pursuit of stable geopolitical relations. International trade brings with it mutual dependencies that may contribute to peace, but at the same time climate change, water, and food shortages are major causes of poverty, political instability, conflict and refugeer outflow in many parts of the world (Brauch and Scheffran 2012; Huntjens et al. 2018). A study by the World Food Programme (WFP 2017) found that the greatest refugee outflows are from countries not only experiencing armed conflict but also the highest level of food insecurity (WFP 2017).

Above developments show there is much more to food than only production, kilograms, and certification. Also legitimation, human security, social justice, crossovers with other sectors such as water, energy, IT, healthcare, and well-being, as well as the pursuit of a healthy and sustainable environment have become increasingly important. A societal transformation towards a Natural Social Contract cannot be accomplished without a transition to a sustainable and healthy agri-food system. There is an urgent need for new production strategies and renewable use of raw materials, new revenue models, and innovative entrepreneurship, as well as new forms of governance and management. Strategies that include social justice and equity in the food system, and strategies that increase the use of ecological processes rather than relying on external inputs for crop production, as well as strategies that foster regional food (distribution) networks and waste reduction (see also Schipanski et al. 2016), are considered as Transformative Social-Ecological Innovations (TSEIs) for realizing a transformation towards a Natural Social Contract.

In the following sections several projects of my research group within the context of the food transition are briefly highlighted.

6.2 NWA Programme 'Transition to a Sustainable Food System'

Thanks to participation of our research group in the national research programme (NWA) *'Transition to a Sustainable Food System'*, funded by the Netherlands Organisation for Scientific Research (NWO), we will be able to conduct transdisciplinary research on the governance of the food transition for a period of 3 years (2021–2023). The consortium consists of eight universities and four universities of applied sciences, in close collaboration with actors from civil society, government and private sector. Next to senior capacity, the project includes eight postdoc positions with the universities involved, while participation of early career professionals and students (through BSc and MSc projects) will be supported and facilitated.

In the Dutch agri-food sector and beyond, many innovative entrepreneurs, citizens, coalitions, and other parties are already actively engaged in the food transition, for example, in realizing short food supply chains (SFSC), community-supported agriculture (CSA), connecting city and countryside, the protein transition, regenerative, circular and nature-inclusive agriculture (NIL), and true cost accounting (TCA). There is a growing support and need for an integrated approach for specific areas, i.e. integrated area development that connects a number of sustainability challenges from a systemic perspective. This requires collaboration with regionally organized networks that connect to area-specific features such as landscape, biodiversity, cultural identity, and social connections within a region. Specific challenges such as nitrogen deposition and climate challenge also require such an area-oriented and integrated approach. The Dutch Ministry of Agriculture's programme 'Innovation on the farmyard' is also in line with area-oriented working. It is therefore obvious to approach the food transition challenge not only nationally but also regionally, in line with the many inspiring (often bottom-up) sustainability initiatives.

An important starting point of this project is to establish links with professional parties and innovative initiatives from practice. This will result in a collective learning process that is highly transdisciplinary in nature. In doing so, we not only involve the 'new' players and initiatives (niches), but also explicitly make connections with large and existing parties (regime), including food producers and supermarket chains that are part of the Dutch food system. This is to gain insight into how their activities may contribute to, or limit, the transition to a sustainable food system. Many of the technologies and governance approaches that we can use for a sustainable food system are known, but how we can further use these insights for a transition to a sustainable food system is still unclear. We want to provide this clarity through a systemic, transdisciplinary approach. The Netherlands is in a good position to investigate and stimulate the transition towards a sustainable food system.

The overarching goal of this programme is to achieve a better understanding of the Dutch food system (as embedded in an international food system), a new design for a more sustainable future food system, and the identification and validation of steering mechanisms and governance approaches to facilitate the transition to such a food system. The central research question is: what is a sustainable food system and which steering mechanisms and governance approaches can accelerate a transition to a sustainable food system in the Netherlands?

We focus on the following sub-questions (with reference to work packages (WPs)):

1. How can the current Dutch food system be defined and delineated: What are the dynamics and what are the interactions of the current food system in the Netherlands? What are the interfaces between the food system and adjacent systems such as water or energy? (WP1 in collaboration with WP4).
2. How can a future sustainable Dutch food system (with possible subsystems) be defined? What new value systems underlie this; based on what characteristics

(e.g. on the basis of widely supported sustainability criteria such as the Paris climate agreement, Convention on Biological Diversity, and SDGs) is their success evaluated? (WP2, in collaboration with all WPs).
3. How does the transition challenge take shape in an area-oriented approach? How do the many sustainability initiatives in the Netherlands actually contribute to the transition to a sustainable food system in the Netherlands and abroad? (WP3 in collaboration with WP2, WP4, and WP5).
4. What is the transition challenge from the current to the new system and which governance arrangements can accelerate this transition? Where are opportunities (e.g. business cases, scaling up excellent initiatives), which obstacles must be overcome (e.g. vested interests, perverse incentives such as agricultural subsidies), where are paradigmatic tensions, the winners and losers of transition? (WP5, in collaboration with all other WPs).

6.3   Nature-Inclusive and Regenerative Agriculture

Most cropland around the world is characterized by large monocultures, whose productivity is maintained through a strong reliance on costly tillage, external fertilizers, and pesticides (Schipanski et al. 2016). Nature-inclusive agriculture (often also called landscape-inclusive or regenerative agriculture) has the potential to offer a more balanced 'production' of food (such as dairy, meat or vegetables and fruit), biodiversity, water quality, carbon storage, landscape, and other ecosystem services than conventional agriculture. In addition, nature-inclusive agriculture may be more resilient to weather extremes (Erisman et al. 2014; Sanders et al. 2015; Van Doorn et al. 2016) and thus more climate-adaptive. A switch to nature-inclusive does require an extensification and substantial adjustments in business operations (Runhaar et al. 2020) and requires an action perspective for farmers in which a transition to nature-inclusive is feasible, also economically. Many farmers have already taken several measures in recent years for more nature-inclusive management, such as extra grazing, adjustment of fertilization, fitting in field edges or green manure. However, far-reaching integration in business operations is still lacking at most companies (Bouma et al. 2020). In addition to compensation and financial incentives, the motivation of farmers is important: motivated farmers dare to take risks and form their own vision on agriculture. Gaining knowledge and experience with nature-inclusive agriculture can play a major role in this. In addition, researchers see that farmers and other stakeholders need an integrated approach to NIL, circular agriculture, and the underlying themes such as climate, soil, water, and biodiversity (e.g. Cuperus et al. 2019; Spoelstra and van Doorn 2019).

Nature-inclusive agriculture starts with healthy soil, produces food within the boundaries of the natural and social environment and has positive effects on biodiversity and the climate (Erisman et al. 2017). A variety of measure can be deployed, depending on type of farming, biophysical, and geographical context. For crop agriculture measures could include non-inversion tillage, field extensions, green manures, catch crops, all year round green fields, reduction of pesticides, flowering

6.3 Nature-Inclusive and Regenerative Agriculture

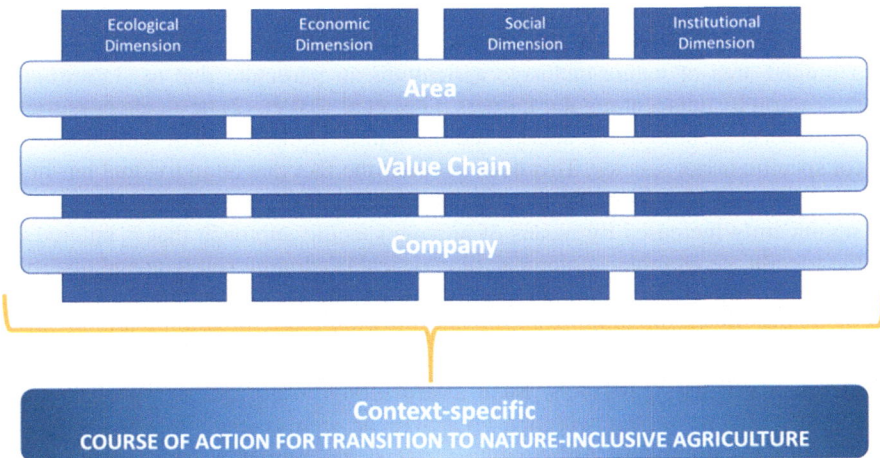

Fig. 6.1 Conceptual model, based on the dimensions of a Natural Social Contract at different levels of scale, for studying a transition towards nature or landscape-inclusive agriculture or circular agriculture

field borders, and landscape elements such as wooded banks and hedges. For dairy farming it could include herb-rich grassland, outdoor grazing, construction of a puddle/wetland system for meadow birds, other cattle breeds in wetland peat areas, fully grass-fed farms, and various landscape elements (for an overview see Erisman et al. 2017).

Nature-inclusive agriculture is not an unambiguous concept, with its origins in the Netherlands, while the term 'regenerative agriculture' is more common in international literature (Rhodes 2017). In general, both concepts entail the inclusion of natural processes into the farming process, while letting nature benefit from the agricultural processes (Degenaar 2019). Since the concept is not yet crystallized, Runhaar et al. (2017) argues 'it can be valuable as a "boundary concept", that brings together farmers, stakeholders and policymakers in order to discuss and negotiate shared meanings and objectives that may contribute to agricultural transformation' (see also Velten et al. 2015).

Exploratory research conducted by my research group provides some preliminary but relevant insights on which factors determine the success and/or failure of a transition to nature-inclusive agriculture in The Netherlands. The following conceptual model, based on the dimensions of a Natural Social Contract at different levels of scale, was developed to guide the analysis (see Fig. 6.1). A follow-up study on how to accelerate the transition to nature or landscape-inclusive and circular agriculture in the Netherlands is scheduled in collaboration with a transdisciplinary research consortium in the course of 2021–2023.

Based on the results of this exploratory research, drawing from interviews (in 2019) with various stakeholders working with the concept of nature-inclusive agriculture in practice, the following key findings can be shared here:

- **Ecological dimension**: an area-oriented approach (transcending individual or smaller farms) is important, because sustainability gains remain sub-optimal for a limited area of nature-inclusive agriculture, in particular due to (1) nitrogen deposition by adjacent regular farmers (Gies et al. 2019), (2) pesticide or nutrient outflow via surface water, and (3) the minimum required area of natural habitat for specific target species. This does not alter the fact that an individual business approach can already yield substantial sustainability gains, but these can be significantly increased by means of an area-specific approach that is tailored to the specific geographic, ecological, biophysical (including soil type and quality), landscape, social, and cultural context of that area.
- **Economic dimension**: an area-oriented approach offers more possibilities for a sustainable and profitable business model, through a system of stacked rewards, for the participating farmers. Such a structural reward system for ecosystem services can be set up within a collective or area council in which various stakeholders (including participating farmers, governments, banks, and nature management organizations) are brought together. The ecosystem services (including more biodiversity, better water quality, carbon storage, landscape elements, etc.) supported by the participating farmers can then be rewarded by, for example, an interest discount at the bank, tax reduction at the water board, a surcharge on the milk price, compensations for CO_2 offsets and easier access to land that would otherwise be unable to be used, thanks to the participation of land and nature management organizations. This is illustrated by the case-study 'Land of Values', where an area council was established in which various stakeholders (i.e. governments, private sector, and NGOs) are brought together to allow a system of stacked rewards for the participating farmers.

  Transforming the farming process and related business model takes time. At the same time, farmers must be able to earn a living. Ensuring that natural and landscape values thrive may be at odds with conventional agricultural business practices, and it is important that the business has enough time to re-organize. This calls for a sustainable revenue model for the next 10–15 years, taking business risks into account. A revenue model with a mix of rewards and rewarding parties, as used in the 'Land of Values' case study, reduces the dependence on subsidies, and financial risks are mitigated by spreading the risks. In this vein, 'new incentives are needed that reward farmers who minimize their ecological impacts, maximize positive impacts, or who switch to biological pest control or use other types of natural processes' (cf. Runhaar 2017).
- **Social Dimension**: a collective with an area-oriented approach offers a social context and legitimacy, making it easier for farmers to switch to nature-inclusive agriculture, instead of an individual business approach. This is because social cohesion and shared values among farmers are very strong and determine what they do, while the negative social pressure on farmers who want to do differently is large (Westerink et al. 2019). A collective with multiple farmers and societal stakeholders can simultaneously ensure better social embedding, dialogue, and mutual understanding between farmers and citizens. Another advantage of

uniting farmers in a collective (in whatever form) is that it provides extra bargaining power in the cooperation and negotiation with other parties.
- **Institutional Dimension:** an area-oriented approach offers more possibilities for multi-level governance, aiming for a better coherence of measures at different levels, including the business level, at the level of a value chain, at the level of a collective or council for one area (with multiple and divergent stakeholders) and at policy level. Policy instruments and 'positive incentives' (and combinations thereof in so-called 'optimal policy mixes') are essential for the realization of a transition to nature-inclusive agriculture, but this can often only be arranged through optimal coordination between central government, provinces, municipalities and water boards and the strive for an area-oriented and context-specific approach. Another important aspect is the development of a tangible vision and mission for nature-inclusive agriculture in a specific area. It is important to identify and create shared and common interests and values. Procedural fairness and social inclusion, as well as options for creating added value for all involved, are important. Agreement on these ethical and normative aspects is important to keep actor coalitions together during multi-year and effective collaboration processes.

Our investigations, so far, underscore the importance of governance in the transformation towards nature-inclusive agriculture. 'Governance is about how farmers, companies in agri-food chains, banks, governments, NGOs and other stakeholders interact and try to influence each other in order to achieve their objectives' (cf. Termeer et al. 2013). Based on above research, an important success factor is the effective collaboration in a collective in one area, instead of individual farms that modify their farming processes.

6.4 Closing the Gaps Between Citizens, Farmers, and Nature

Recently, interest in local food systems has skyrocketed (Pigford et al. 2018), centred around an approach in which citizens, farmers, and other stakeholders work together to create a sustainable, healthy, and predominantly local food system. Within this context three examples are highlighted:

- **Food policy councils (FPCs)** as loci for practising food democracy, for example, in Germany (Sieveking 2019), the USA (Scherb et al. 2012; Gupta et al. 2018), and Canada (Sussman and Bassarab 2017).
- **Community-supported agriculture (CSA)** with its origins in Europe and North America, and similar to the TeiKei system in Japan (Kondoh 2015), and more recently also initiated in China and Singapore. It is a sustainable alternative to industrial agriculture, in which there is direct producer–consumer transaction (White 2020) that allows the producer and consumer to share the risks of farming (Galt 2013).

- **Short Food Supply Chains (SFSC)**, which 'aim to reconnect the two extremities of the food supply chain, reconcile producers with citizens, stimulate mutual trust and establish a short chain based on common values on food, its origin and production method' (cf. SKIN 2020).

In the Netherlands, more and more citizens want to participate in the transition towards a healthy and sustainable food system (BoerenBusiness 2018). Our research group is currently working to develop a pilot on community food forestry, with the central aim of restoring the natural relation between citizens, food, and nature. A food forest is a multifunctional approach to increase food security and provide ecosystem services (Clark and Nicholas 2013). A well-functioning food forest is actually the most ideal form of circular agriculture, since a healthy ecosystem shows an important basic principle of the Circular Economy: that is, how you organize circularity at the lowest possible level. When combining a food forest with the concept of community-supported agriculture, it provides many opportunities for social cohesion and citizen participation to restore the relationship between citizens, food production, and nature. Community Food Forestry (CFF)/Community Agro-Forestry (CAF) allows citizens to become co-owner of their own food system. A first pilot is being established by Inholland University in the Netherlands. Other examples that are being studied by our research group include:

Rechtstreex in Rotterdam, an example of short food chains and closer ties between citizens and farmers (see Fig. 6.2), Burgerboerderijen (Citizen Farms Cooperative), and Herenboeren (freehold farms). The aim of the Citizen Farms Cooperative is to reconnect people with the source of their food: 'their' farmers. Establishing direct, local connections between citizens and farmers shortens the distance from grain to bread.

Fig. 6.2 Rechtstreex Rotterdam, an example of the short food chain and closer ties between citizens and farmers (Photocredits: Rechtstreex)

In general, these projects aim to build a community that coexists with the natural environment and to reconnect people with the source of their food through mutually supportive relationships between farmers and consumers. This has a great deal of positive effects, such as fairer food, no undesirable increases in scale and the shortest possible food chain, without delegating control to supermarkets (Burgerboerderijen 2019). A similar concept is employed by Herenboeren (freehold farms), which are co-owned by groups of consumers who employ demand-driven production practices and consume what they produce (Herenboeren 2019).

6.5 Measuring Sustainability and Health Aspects of Our Food Chains

Major changes are visible within our current food system: take a look at the shelves of the average Dutch supermarket and compare them to the situation 15 years ago. A lot has changed, and shelves with sustainable, organic, and ecological brands are getting more common. Quality, taste, production methods, and fair prices for farmers are becoming increasingly important for consumers, and organic and biodynamic supermarket chains such as Ekoplaza and Odin are gaining increasingly large market shares (Distrifood Dynamics 2019). Traditional premium brands and supermarket house brands have stepped up their efforts to give their products a greener, healthier, and fairer image. In some cases this might result in greenwashing, without necessarily leading to more sustainable or healthier products.

In the Netherlands alone, there are almost a hundred quality marks and labels for sustainable, fair, and responsible food. Most of these, however, were developed by businesses themselves, so they are effectively marking their own papers. Coca Cola, for instance, does not hesitate to put a green tick on its bottles, even though they are full of sugar. After all, they claim, it contains no salt, fat, and saturated fat. Perhaps, consumers would benefit more from a smaller number of transparent and independent labels, indicating whether a certain product is animal-friendly, organic, or fair-trade, for instance. These labels must be transparent and reliable, which means they must be subject to strict audits, and quality standards preferably should be laid down in law at the national or European level.

An important challenge is making sustainability and health measurable for consumers, government, and businesses. In the field of healthy food, the nutricode is a promising development, a score that can be used to 'calculate' all the healthy and unhealthy properties of a given foodstuff. These scores can be printed on packaging, to indicate how healthy or unhealthy the product in question is, with A being the healthiest and E the least healthy. This system already exists in France and is currently introduced in Spain and the Netherlands.

My research group is currently working with partners on developing a roadmap for measuring, improving, and communicating sustainability in livestock farming from an integrated perspective. For instance, the carbon footprint is an important indicator for measuring the sustainability of livestock. However, the assessment and labelling of emissions is difficult and a decrease of carbon footprint is often at odds with other sustainability criteria such as biodiversity and extensive grazing. There are also other themes of societal importance, such as the relationship between citizen and farmer, animal welfare, landscape, nature, biodiversity, and cultural history. The various aspects of sustainability, and diverse interests, can be found in the various separate quality marks that have been developed. The aim of this project is to form an integral overview of quality marks, measurement methods, and sustainability criteria for livestock farming, consumer perceptions and to provide insight into the areas of tension and diverse interests between them. This information is important to determine the impact that a certain intervention would have on people, animals, and the environment. From the overview, a roadmap is designed for the further development of existing quality marks with regard to criteria, methodology, allocation, in order to match the needs of different target groups, including consumers and business customers. All sectors within meat-producing livestock farming are taken into account.

6.6 South Holland Food Family: Transition Towards a Sustainable and Self-Sufficient Food System

The Dutch province of South Holland, with 3.6 million inhabitants living on 3403 km^2, is one of the world's most densely populated areas. It includes both Rotterdam, Europe's largest port, and The Hague, the country's second and third-largest cities. Remarkably, the province has a large agricultural sector, with arable farming, bulb farming, livestock farming, and even the world's largest contiguous greenhouse area. However, most South Holland food produce is exported—the province's level of self-sufficiency is currently approximately 40% (Nefs 2017).

The South Holland Food Family (in Dutch: Zuid-Hollandse Voedselfamilie) is an open innovation and food transition network, supported by the provincial government and many partners. The ambition of the Province of South Holland is: 'More sustainable agricultural and food chains, offering healthy, sustainable and affordable food for everyone in the Province of South Holland in five to ten years from now'. Part of this ambition is to achieve a provincial level of 80% self-sufficiency in 2036. That would save a lot of food miles and yields even fresher products. Moreover, it would strengthen the bond between farmers and citizens, while at the same time, increasing more citizen awareness on the production process itself. But above all, the ambition is to realize a more sustainable food system.

This ambition cannot be achieved through improvement of the current food system. Rather a transition is needed—a fundamental change of the food system's structure, culture, and practice. The Province has adopted a transition approach in its 2016 Innovation Agenda for Sustainable Agriculture. This approach adopts elements from transition management (Loorbach et al. 2017), technological innovation systems (Hekkert and Ossebaard 2010), and a 'Networked Working'-approach (Wielinga and Robijn 2018). Internally, government workers call this combination the 'change approach' and its main goal is to stimulate and facilitate experimentation, innovation, and entrepreneurship within the food transition.

Organizationally, the transition approach entails: (1) the aforementioned open innovation network for food pioneers and change makers called the South Holland Food Family, (2) a subsidy programme to support experimental projects for a sustainable food system, which has initiated an impressive portfolio of more than 30 experimental projects (in Dutch: Proeftuinen), where food pioneers and change makers show what changes are possible, and (3) a knowledge and development programme to further develop and disseminate knowledge from the experimental projects, making use of reflexive monitoring, impact assessments, and a dynamic learning agenda. The open innovation trajectory followed by the South Holland Food Family is visualized in Fig. 6.3.

The TSEI analytical framework (Sect. 5.1) has been used to analyse institutional change during initiation, development and implementation during the first three years (2015–2018) of the South Holland Food Family innovation network (Huntjens et al. 2020). The framework was used to zoom in on a series or cluster of related action situations (and their context), looking at 'structure' and 'agency', and at the output-outcomes-impact of these action situations (per situation where possible and per series/cluster). A series or cluster of related action situations is referred to as an action arena or transition arena. An important first step in applying the TSEI-framework, in this case, was a timeline analysis, because it aims to provide a better insight into a series of action situations and the associated learning history. The timeline method is thus part of the methodology of the TSEI-framework. A total of eight action situations were selected in which we could observe an informal or formal steering of the process (see Table 6.1), based on empirical data from a series of individual interviews with participants of these actions situations, and based on joint reflection on the process during a timeline session with multiple participants. The informal and formal steering and related institutional change that was observed differs per action situation, but usually involves a situation where multiple parties (with different interests, perspectives, and preferences) come together and are confronted with a series of potential actions, where these parties exchange goods and services, try to solve problems, influence each other, learn from each other, and resulting in shared output/outcomes. Table 6.1 provides a brief description of the nature of these action situations and the formal or informal institutional change that occurred.

The example of TSEI-framework application provided here shows when and how local agents change the institutional context itself, which provides relevant insights on institutional work (Beunen and Patterson 2019) and the mutually constitutive nature of structure and agency (e.g. Giddens 1984; Bourdieu 1988, 2005; Seo and Creed 2002). Above institutional analysis also shows the pivotal role of a number of actors, such as network facilitators and provincial deputy, and their capability and skills to combine formal and informal institutional environments and logics and mobilize resources, thereby legitimizing and supporting the change effort. The results are indicative of the importance of institutional structures as both facilitating (i.e., the province's policies) and limiting (e.g. land ownership) transition dynamics. Interestingly, while the provincial government holds some power over such institutions, it also has to operate in wider national and EU- institutional settings

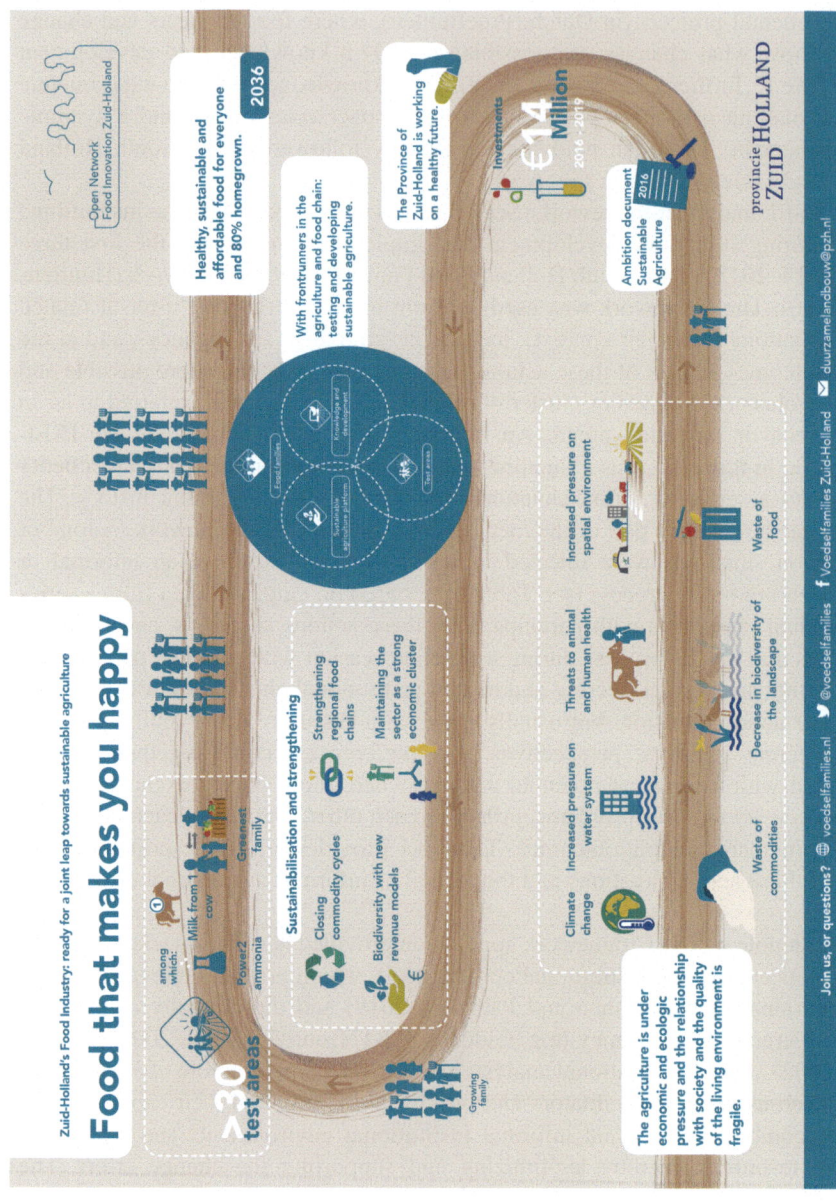

Fig. 6.3 Transition path of the South Holland Food Family

Table 6.1 Example of TSEI-framework application: Overview of action situations where institutional change occurred, during the initiation and development (2015–2018) of an open innovation and food transition network called The South Holland Food Family

	Name and date of action situation	Type of formal or informal institutional change that occurred
1	24-h team creation meeting on October 8 & 9, 2015	This meeting introduced the open innovation network approach to various stakeholders (including Province) in the South Holland agricultural sector, and formed the basis for further network meetings. Using the TSEI-framework we can observe some clear examples of process steering in this action situation: First, there is a deliberate choice to use the multi-level perspective (MLP) from Transition Management theory, and as a result, to search for front-runners instead of established or 'regime-confirming' parties. Also the Free Actor Network (FAN) approach was used as a reference for (informal) network formation. From the agency side: the interviews show that the informal setting played an important role in distancing the civil servants from their official role, and also for other attendees, to be able to speak freely about the food system (opportunities, obstacles, and ambitions) and find their own role. The central question was: who wants what, who can do what, and who has influence? This formed an important basis for the network (yet to be established) and also for determining the transition agenda. Outputs: (1) 'The book of ideas—Baseline for more sustainable agriculture'; (2) A joint problem definition and way of solution thinking; (3) Concrete plans for further steps: interviews with front runners; energy among participants. Relatively fast follow-up also gave participants confidence that it might work this time.
2	Kick-off meeting at Duijvestijn, entitled 'Who is following the madman?' on January 14, 2016	The network comes together for the first time. The structure-agency dynamics can be observed very nicely in this action situation. For instance, various forms of stakeholder management and process facilitation take place: (1) The Province has given a 'go ahead' to set up an innovation platform, with an active role and commitment of the business community. The first meeting (the kick-off) was therefore at Duijvestijn's company, one of largest growers of tomatoes in the Netherlands; (2) The presence of the Provincial Deputy (Han Weber) is an example of latent or implicit power from a TSEI perspective: so power is not used explicitly, but others do take into account the power that can be (or will be) exercised; (3) Management by objectives (such as working towards the first progress report) and more general agreements; (4) Process steering by interactive games and related game rules, with the aim to create energy. Everyone is invited to play, a lot of innovators are invited and interviews are held to bring their ideas. In short, it was deliberately set up as an open network, so anyone who was interested could join and was invited to contribute.
3	Network meeting at Blue City on 28 January 2016	Second network meeting. During this meeting the term 'Food Families' was created, which was considered an important moment by a number of those involved. The choice of name and elaboration is a form of process steering, because the network initiators aim for a family feeling, where the members feel connected to each other. And that is what the initiators want in the food chain. Ideas are sought that give energy.

(continued)

Table 6.1 (continued)

	Name and date of action situation	Type of formal or informal institutional change that occurred
4	Network meeting at Corné van Leeuwen on April 21, 2016	An early version of the Transition Agenda is shared and the first progress report is almost completed. During this meeting three farmers take centre stage and visions of the future are formulated. Everyone realizes: it starts with them.
5	Food family breakfast at the Province of South Holland on May 18, 2016	Results of earlier network meetings are presented in the form of a transition agenda, as part of the first progress report. Three possible transition paths are presented here. This meeting was instrumental for the final adoption of the Ambition document on the Innovation Agenda for Sustainable Agriculture by the Provincial Council of Zuid-Holland on 26 May 2016, including the allocation of funding for the innovation network, subsidized pilot projects (>30), and a knowledge team. In the meeting of 18 May, provincial decision-making and politics had a central place, which is a prime example of where the formal decision-making process (by Province) and the informal decision-making process (by innovation network) coincide.
6	Food innovators party at Koppert-Cress on 11 May 2017	A large gathering that made the Food Families visible and created a lot of energy in the network. There were workshops, inspiration sessions and various ideas were presented and linked again, leading to a tightening of the agenda (output) and consolidation of the network (output).
7	Taste makers meeting on 18 & 19 September 2018	After 2 years of Food Family, it was time for an evaluation, and that question was put out by the Province in the form of a 2-day multi-stakeholder meeting and interviews, and aimed to gauge the state of the energy in the network and contributed to the feeling of energy in the network. Output: the report 'Taste makers thermometer' (2018)
8	Harvest Day on October 18, 2018	The Harvest Day was a major event, generated energy and opened a 'window of opportunity' to continue. In parallel workshops, current problems in the pilot projects were tackled, of which relevant examples for the TSEI analytical component structure/institutions included the bureaucracy around POP3 (subsidy) processes, and legislation and regulations that appeared to block short supply chain (SSC) projects. Output: During the Harvest Day, it was determined what is needed to be able to continue in South Holland and a campaign plan 'Manifesto of the Food Families' was handed over to provincial and national decision-makers. This is another example of where the formal decision-making process (by Province) and the informal decision-making process (by innovation network) coincide.

6.6 South Holland Food Family: Transition Towards a Sustainable and...

that are beyond its direct influence. This changes the role of the province. Where it started out as an "enlightened incumbent" with an innovation programme, it now is slowly taking on a more 'pioneering' role in its wider institutional environment. Hence, the transition policies reflexively have changed the province's role and identity. More details about the application of the TSEI-framework in this case-study can be found in a conference paper by Huntjens et al. (2020).

Open Access This chapter is licensed under the terms of the Creative Commons Attribution 4.0 International License (http://creativecommons.org/licenses/by/4.0/), which permits use, sharing, adaptation, distribution and reproduction in any medium or format, as long as you give appropriate credit to the original author(s) and the source, provide a link to the Creative Commons license and indicate if changes were made.

The images or other third party material in this chapter are included in the chapter's Creative Commons license, unless indicated otherwise in a credit line to the material. If material is not included in the chapter's Creative Commons license and your intended use is not permitted by statutory regulation or exceeds the permitted use, you will need to obtain permission directly from the copyright holder.

Governance of Urban Sustainability Transitions

My research group is involved in collaborations with the dynamic 'Amsterdam Metropolitan Region (MRA)' and 'Rotterdam-The Hague Metropolitan Region (MRDH)', with the objective to investigate the complex governance challenges and opportunities related to urban sustainability transitions, mainly through transdisciplinary collaboration. The resulting knowledge and skills are used to support and engage with Transformative Social-Ecological Innovation (TSEI) in-the-making, which in turn will generate new knowledge and skills (i.e. in iterative learning cycles). This chapter starts with a brief overview of urban sustainability challenges (Sect. 7.1). Research activities are centred around the transition to climate-resilient and healthy cities (Sect. 7.2), feeding and greening megacities (Sect. 7.3), as well as the transition from linear to circular and regenerative economies and cultures in (mega) cities (Sect. 7.4). In parallel, a new transdisciplinary Minor is developed, called 'Collaboration for the City of the Future' (Sect. 7.4).

7.1 Urban Challenges and Developments

Sustainable Development Goal 11 (SDG 11), in particular, is dedicated to making cities inclusive, safe, sustainable, and resilient, as part of the UN Agenda for 2030, while cities play a central role in reaching many of the other SDGs. However, the cities of the twenty-first century are facing enormous challenges in reaching the required level of sustainability and improving the living standard of citizens. More and more people are living together in an increasingly compact space. They are looking for a better life, employment, better education, better healthcare, better infrastructure, and social services.

Urban population is expected to grow with 1.5 billion in the next 15 years, and 3 billion by 2050 (cf. United Nations, World Urbanization Prospects 2018). 'Cities today occupy approximately 4% of the total land, but contribute 70% of global (GDP), over 60% of global energy consumption, 70% of greenhouse gas emissions, and 70% of global waste' (ibid.). How the world meets the challenge of sustainable development, in particular for reaching many of the Sustainable Development Goals (SDGs), will be intimately tied to the process of urbanization and the governance of urban sustainability transitions. Hence, cities need to be organized more efficiently and sustainably with regard to the supply of food, raw materials, and energy, as well as finding ways of adapting cities to climate change, keep the cities attractive and healthy, and to address poverty and inequalities (see below overview of urban challenges).

While urbanization poses serious challenges, cities are also breeding grounds for innovation and can be powerhouses for sustainable development, if the right policies are put in place. 'Cities provide a wealth of opportunities, jobs included, and generate over 80% of gross national product across the globe' (cf. UNFPA 2019). 'Nowhere in human culture is the centrality of collaboration and sharing more obvious than in the city' (cf. Agyeman and McLaren 2017). The city is not just a venue for sharing, but is historically a shared entity in itself: the result of shared co-production (ibid.). For many cities, though, the pace of urbanization is overwhelming both national and local capacities to capitalize on the opportunities before them (C40 2018). 'The problems are multi-sectoral, the available data is incomplete, and there is a lot of disagreement over how to proceed. The most common challenges include unplanned urban expansion, ineffectual governance and legal frameworks, and a dearth of local-level revenue generation mechanisms' (cf. C40 2018). 'Cities with the greatest infrastructure needs often lack the capacity and knowledge to develop bankable projects. This is exacerbated by limited access to credit and an insufficient ability to take advantage of endogenous sources of finance, which, for example, could be used to invest in core infrastructure such as water, drainage, and energy' (ibid.).

Within this context, several important urban sustainability challenges can be identified (in no particular order):

- **Population growth**: Projections are that the world population will grow from 7.8 billion in 2020 to 9 billion in 2037 (World Population Prospects: The 2019 Revision). Approximately 55% of the world's population lives in urban areas, a proportion that is expected to increase to 68% by 2050 (ibid.). Projections are that by 2030 there will be 41 megacities, defined as urban agglomerations of more than ten million people (United Nations 2018). 'While population is expected to continue growing exponentially across most of the globe, this is less so in Europe. While many challenges still faced are related to population pressure, Europe also has to cope with new challenges related to a declining and ageing population in many cities' (cf. JRC 2019).
- **Climate change**: Cities account for more than 70% of global energy use and related CO_2 emissions, and thus play a key role in climate change mitigation (IPCC 2014; UN Department of Economic and Social Affairs Population Division 2015; Hopkins et al. 2016). Examples of climate mitigation can be found in the ongoing urban energy transition (e.g. solar panels, recycling and waste management, high-performance insulation, etc.), and in transportation (e.g. electric vehicle fleets for cities, autonomous vehicles, microtransit, green

spaces for improving air quality, mitigation of the urban heat island effect, etc.). At the same time, cities are vulnerable for the impacts of climate change, with increased risk of heat-related deaths, extreme weather events, food and resource shortages, power outages, and infrastructure failures. 'Over 90% of all urban areas are coastal, putting most cities on Earth at risk of flooding from rising sea levels and powerful storms' (cf. The Global Risks Report 2020; World Economic Forum 2020). One of the major challenges for the governance of future cities is the combination of climate change and the local urban heat island effect (Mohajerani et al. 2017; UN-Habitat 2020). Hence, climate change adaptation is necessary for creating more resilient and healthy cities, for example, by means of green-blue infrastructure (see Sect. 7.2).

- **Feeding megacities**: 'Urbanization will drive intensified consumer demand and value chain concentration, while the distance between food producers and food consumers will continue to widen' (cf. NewForesight and Commonland 2017). For more details, see Sect. 7.3.
- **Mobility:** 'Environmental pollution, congestion, and long commuting times are just some of the issues related to mobility in cities. A decrease in ownership of private vehicles in favour of efficient and connected public transport and active mobility modes could greatly ease these problems' (cf. JRC 2019).
- **Affordable housing**: 'The recent scale-up of foreign and corporate investments in residential urban property has transformed patterns of ownership. Prices are recovering faster than earnings, and the availability of housing is low. Short-term rental platforms may also cause property prices to spiral and negatively affect local liveability' (cf. JRC 2019).
- **Poverty and inequalities:** 'Cities are home to high concentrations of poverty. Nowhere is the rise of inequality clearer than in urban areas, where wealthy communities coexist alongside, and separate from, slums and informal settlements' (cf. UNFPA 2019). 'Many people are also likely to be pushed into poverty due to higher prices of essential commodities in urban areas' (cf. WHO 2016). In addition, more enduring patterns of inequality need to be taken into account (Tilly 1998; Tonkiss 2017). Some of the deepest and most persistent patterns do not only derive from economic factors, but are generated around social distinctions which legitimize the unequal distribution of resources and opportunities between different groups across different contexts (Tilly 1998; Tonkiss 2017), which in turn could result in economic differences. Therborn and Aboim (2014) conceptualized multiple forms of inequality: (1) vital inequality (differential health outcomes, mortality rates, life expectancies, distributions of hunger and malnutrition, exposure to environmental and other types of somatic risk), (2) resource inequality (access to and command over economic and non-economic resources, goods, or capitals), and (3) existential inequality (disparities of dignity, autonomy, freedom, opportunity, and self-determination). From a sociological perspective these kinds of inequalities are treated as a relational problem, in terms of dynamic social relations between individual and groups (Tonkiss 2017).

- **Urban health:** The types of urban health challenges are varied. Non-communicable diseases (NCDs)—the diseases that result from a combination of our biology, how we live our lives, and the environment we live in—are endemic to city life (WHO 2016). A study of more than 100 countries found that body mass index (BMI) and blood cholesterol levels, both major risk factors for NCDs, rose rapidly with increases in national income and level of urbanization (Ezzati et al. 2005). WHO estimates that 68% of global deaths were caused by NCDs in 2012 (16). Much of this burden will be concentrated in cities. In addition, urban environments offer favourable grounds for the spread of infectious diseases, especially in areas of high population densities with low resources such as slums. Increased international travel and migration have resulted in cities becoming important hubs for the transmission of infectious diseases, as shown by recent pandemics such as H1N1, Ebola virus, and Corona-virus (COVID-19). Other urban health concerns include air pollution and mental health issues, while the concentration of poverty in overcrowded urban areas also constitutes an increased risk for violence and injuries (WHO 2016). Emerging trends, such as ageing, and the prevalence of malnutrition and obesity and mental health in cities have to be tackled with a long-term effort (JRC 2019).
- **Environmental footprint**: 'Providing water, energy, and food security for urban populations results in significant environmental pressure beyond city boundaries. Four of nine planetary boundaries have already been exceeded due to human activities. Several lifestyle and behavioural changes can help city inhabitants significantly reduce their environmental footprint, such as shifting to a healthy diet, reducing waste, using active or public mobility modes, or choosing sustainable energy sources' (cf. JRC 2019).

In short, cities are facing enormous development challenges, such as climate change adaptation, the energy transition (i.e. climate change mitigation), circularization, preservation of biodiversity, improving quality of life, addressing poverty and inequalities, and maintaining and improving public health (Gallopín 2006; Gill et al. 2007; Pötz and Bleuzé 2016; Peek 2015; Huntjens 2019). These metropolitan challenges are often interrelated and will have to be approached from a complex system perspective (see Sect. 4.6). It will require new forms of urban governance, management, and planning that are able to deal with complexity and uncertainty (see Sect. 4.7). There is no 'one-size-fits-all' recipe. These challenges are particularly complex because of the multitude of actors involved in the process. Actors such as urban planners, environmental lawyers, spatial planners, water boards, energy experts, green space managers, urban farmers, citizens, and businesses will have to work together on the technological, legal, financial, and administrative aspects of this sustainability transition. All of whom have their own interests and often divergent perspectives and problems and solutions, complicated by the often sector-specific distribution of financial budgets, and the fact that existing laws and legislation in many instances provide insufficient scope for climate-proof or circular solutions. In the Netherlands, for example, central government is increasingly transferring responsibility for the sustainability transition to municipalities, companies, local organizations, and citizens. While decentralization and innovative planning processes with intensive participation are considered necessary for urban governance, it is often a new and overwhelming challenge for these parties to jointly, efficiently, and effectively transform the city into a future-proof living environment.

Research by Prendeville et al. (2018) and Fratini et al. (2019) highlight various shortcomings in the governance of urban sustainability transitions in European cities:

- *'limited attention for understanding and redirecting existing consumption patterns;*
- *lack of attention for methods of participatory processes for co-production of urban circularity, and little emphasis is given to the inclusion of citizens and communities;*
- *too much emphasis is given to major incumbent actors, in particular to the role of business and smart, digital, and data-driven technologies'* (cf. Fratini et al. 2019).

To make the urban sustainability transition a success, there is a need for new, efficient, and inclusive forms of urban governance, organization, and cooperation between different urban actors and for new decision support and knowledge management tools to support such collaborations. Within this context, making sustainability and the feasibility of possible interventions measurable, or at least create more insight on costs, benefits, risks, and potential trade-offs, helps the parties involved to make better decisions.

In below section I will highlight several examples of applied research activities, where our university together with partners is working on systemic solutions for complex urban sustainability challenges.

7.2 Climate-Resilient and Healthy Cities

'Cities are experiencing multiple impacts from global environmental change, and the degree to which they will need to cope with and adapt to these challenges will continue to increase' (cf. Elmqvist et al. 2019). 'A resilient city provides access to healthy food, clean water and air, safe transportation infrastructure, healthy buildings, and health services *for all* citizens' (cf. Newman et al. 2017). 'The integration of grey, green, and blue infrastructure in urban planning through institutional innovation and structural reorganization of knowledge-action systems may result in large health improvements and increase urban resilience' (cf. Elmqvist et al. 2019).

For instance, cities in Europe, the USA, India, China, Australia, and several other countries are using the concept of a sponge city to design climate-resilient and healthy cities. 'The idea of a sponge city is simple—rather than using concrete to channel away rainwater, you work with nature to absorb, clean, and use the water. Examples include eco-friendly terraces that are used during the dry season as a park for residents to enjoy, while it provides retention capacity during heavy rains, which reduces flooding in cities and prevents disasters and their subsequent costs. It protects the city with less reliance on grey infrastructure like flood walls, dykes, or drainage systems. Not only does this safeguard the city by working with nature, but the water is clean, vegetation can grow, and a habitat is created for improving biodiversity. The abundant vegetation found in urban wetlands acts as a filter for domestic and industrial waste and this contributes to improving water quality. Other measures include green walls and roofs, permeable pavements, and green buildings' (cf. Myers 2019).

An example in the Netherlands is 'The Most Sustainable Square Kilometre', which is an area surrounding Leiden Central Station in the midst of a sustainable transformation into the most sustainable kilometre in the Netherlands by 2025. To realize this ambition, 29 organizations, including governments, residents, the private sector, and knowledge institutes, have signed a Green Deal in 2018. A recurrent governance question is how such a cooperation processes can be organized. This requires a better understanding of policy and strategic decision-making, processes of trust-building and conflict resolution, group decision-making for effective urban planning, and to ensure accountability and legitimacy that guarantees balances of interests and perspectives. The transition to climate-resilient and healthy cities will raise normative questions on what makes cities climate-proof and healthy and who should bear the costs involved in the process (Eriksen et al. 2015). Within this context, the integration of important bottom-up processes of learning and knowledge development with top-down policies and strategies is an important challenge.

7.3 Feeding and Greening Megacities

'Urbanization will drive intensified consumer demand and value chain concentration, while the distance between food producers and food consumers will continue to widen' (cf. NewForesight and Commonland 2017). Since 80% of food will be consumed in cities by 2050, cities can significantly influence the way food is grown, particularly by interacting with producers in their peri-urban and rural surroundings (Ellen MacArthur Foundation 2019). However, current global and regional food systems do not consist of closed cycles, in particular related to CO_2, minerals, organic matter, water, and energy. This means that precious building blocks are thrown away. This can lead to scarcity (for example, phosphate), but also to environmental impact (for example, nitrogen and CO_2). 'By valorizing and/or reusing residual flows and minimizing losses, it is possible to contribute to a food supply that has less impact on the environment. Though much of the foodprint is embodied within imported foodstuffs, cities can still implement design and policy interventions, such as improved nutrient recycling and food waste avoidance, to redress the foodprint' (cf. Goldstein et al. 2017). 'A city's foodprint can rise or fall based on several factors, including the kind of food eaten (grain-fed versus grass-fed meat, meat versus vegetables, water-intensive versus less thirsty crops, etc.), the amount of food wastage, the distance food travels to the city, and other factors' (cf. Gardner et al. 2016).

A potential major contributor to progress in feeding global megacities involves reevaluating regional food systems so that each is much more efficient and self-sustaining than in the current global supply chain (see example of the South Holland

Food Family in Sect. 6.6). One major advantage is that food production will be closer to the people eating it, resulting in a lower carbon footprint since there will not be long, and often global, cycles of logistics. However, transport is just one factor in the total carbon count of food, while the carbon footprint is only one metric of sustainability, besides social, economic, and other environmental dimensions (Morgan 2009). Therefore, the sustainability of local food systems needs to be measured across multiple dimensions. For example, one important advantage of local food systems is a smaller divide between farmers, consumers, and nature (Sect. 6.4). Hydroponic systems and vertical farming in urban areas also get a lot of attention, although the extent to which they can feed megacities is yet unclear, and these can only be partial solutions for high-yields and ability to supply vast dietary needs of city residents. Section 7.3 will address the potential of circular and regenerative cities as an important development for feeding and greening megacities.

7.4 From Linear to Circular and Regenerative Cities

Circular Economy and Regenerative Economy are major themes in urban development, with the aim to be smarter and more efficient with energy, resources, and waste in order to prevent further depletion of raw materials and growing landfills (Kirchherr et al. 2017). "The concept of 'regenerative cities' is seeking to address the relationship between cities and their hinterland, and beyond that with the more distant territories that supply them with water, food, timber, and other vital resources" (cf. Girardet 2017).

Cities today are huge consumers of resources in the purest style of linear economy: make, use, and disposal of resources. As a counter-proposal, is it possible to create future cities that regenerate as many resources as they consume? In any case, 'circular flows of materials in cities promise job creation, operational savings, less waste, and lower carbon emissions' (cf. C40 Cities 2018). Even more so, in a regenerative city, all organic waste is reused, all other materials are separated, recycled, or upcycled, while for some cities all energy needs to come from the sun and wind (e.g. Adelaide), while other cities rely more on hydropower (e.g. Basel). According to UN-Habitat (2020), 'a regenerative city would benefit the environment and the natural ecosystems, driving local economy, and improving the connection and cultural life of its neighbourhoods. Such a city would guarantee its capability for a constant and automatic renovation becoming a vector of prosperity and an essential tool to attain the Sustainable Development Goals (SDG) and to combat climate change' (cf. UN-Habitat 2020; Schmitt et al. 2019).

Does that sound utopian? On the contrary, many inspiring examples of cities in transition to circular and regenerative cities, with a vision to become an 'Eco-City' or 'Ecopolis', can be found around the world (see below examples of Tianjin, Adelaide, and München).

- **Tianjin Eco-City in China**: This Eco-City, expected to be completed in 2020, will make use of the latest sustainable technologies such as solar power, wind

power, rainwater recycling, and wastewater treatment/desalination of sea water. In order to reduce the city's carbon emissions, residents will be encouraged to use an advanced light rail system, and China has also pledged that 90% of traffic within the city will be public transport. The city includes varied eco-landscapes ranging from a sun-powered solarscape to a greenery-clad earthscape for its estimated 350,000 residents to enjoy. In addition to these typical sustainability goals, social harmony is promised to be a key feature of this Eco-city, for example, through subsidized public housing for lower and lower-middle income groups.
- **Adelaide in Australia** has started with an ambitious plan 15 years ago to become a circular and regenerative city. The city of 1.5 million inhabitants now runs for 45% on wind and solar energy; storage takes place in a 100 megawatt battery; CO_2 emissions have fallen by 15% since 2000; all organic waste is composted and reused by farmers north of the city; a large-scale reforestation programme ensures air purification; thousands of new green jobs have been created; local democracy is nourished, administrators have a long-term vision again.
- **München (Germany):** 'The municipal utility company (Stadtwerke München) aims to supply every customer with renewable energy by 2025, reduce CO_2 emissions by 50% by 2030, and become the first German city to have District Heating that relies solely on renewable sources by 2040' (cf. Ecopolis 2020).

In the Netherlands, the national government has decided that the Netherlands should be circular by 2050. The energy transition is currently in full swing and topics such as healthy cities, fighting social inequality, and strengthening cohesion in culturally diverse neighbourhoods have enjoyed a steady rise in popularity, as well as new ways to do business, such as social entrepreneurship. Municipalities are looking for new ways to embrace and respond to initiatives by citizens and businesses aimed at sustainable development. Water boards in the Netherlands, for instance, are working on producing energy and raw materials from wastewater, transforming waste and water treatment systems into energy and resource plants. In general, green-blue grids are considered one of the structural building blocks of circular cities, as well as natural ways for closing urban loops (Pötz and Bleuzé 2016). In general, making cities greener and bluer contributes to climate adaptation (see Sect. 7.2), improved biodiversity, improved air quality, energy production from surface water, wastewater, and biomass, recovery of raw materials, such as phosphate and nitrogen from wastewater, boosts quality of life and health, space for recreation and slow traffic, and provides scope for stakeholder participation and citizen engagement (ibid.).

But how does urban planning and governance towards an Eco-City or Ecopolis work in practice, and what urban governance arrangements are required to catalyse a transition from a linear economy to a circular and regenerative economy? C40, a network of the world's megacities committed to addressing climate change, has researched 40 in-depth case studies from cities around the world, demonstrating how municipalities can advance towards zero waste economies (C40 2018). An important finding is that city municipalities need to break away from the traditional configuration of government ministries and administrations working in silos (and with sectoral budgets), and to develop common and integrated work programmes and budgets. It requires integrated area development that connects a number of sustainability and social challenges from a systemic perspective. In addition, multi-stakeholder governance and approaches to balance top-down policies and visions with important bottom-up stakeholder processes are considered as one of the most important challenges for translating circular economy principles (see Table 7.1) into urban planning and governance.

One of the major findings of the Future Cities-report (UN-Habitat 2020) is that future cities need citizen-centred governance, where the inhabitant will change from being the subject of observation to become an active partner in the governance of the city. Urban planning and design could be reorganized in such a way that more problems can be solved at the community level (the subsidiarity principle), with stronger citizen engagement and by new coalitions in horizontal innovation networks (see Sect. 3.7). 'Also smart city technology will enable the individual citizen and community of citizens to plan, build, and govern more sustainable, resilient, and regenerative cities' (cf. UN-habitat 2020). In any case, a regenerative city requires a strong sense of community and respect towards their fellow citizens and towards the environment (UN-habitat 2020), which closely resembles the contours of a Natural Social Contract (Sect. 3.7).

The transition to a circular and regenerative city will take place in different sectors (e.g. construction and buildings, agri-food, packaging, electronics, etc.), and at different geographical scales on which cycles can be closed (i.e. local, regional, or international), but will also depend on issues related to ownership, user rights, and multiple value creation. New forms of value creation and inter-organizational circular business models are needed for closing the loop of a product life cycle, and it is unavoidable that questions will arise on whether business capital should be private property or property of the community or a business collective, and how this could be (re-)organized (see Sect. 4.4). 'There is a growing number of urban commons showing that it is not only possible but highly attractive to create commons through which citizens can actively participate in the circular design of their food system, housing or city spaces and the programmes and policies that govern them' (cf. Bollier and Helfrich 2015). 'Several governance benefits are associated with urban green commons such as cost reduction for the management of urban green spaces, as well as designs for reconnecting citizens with nature. Urban green commons play a key role in transforming cities toward more socially and ecologically benign environments' (cf. Colding et al. 2013).

7.5 Collaboration for the City of the Future

The present societal challenges require a transition to more transdisciplinary collaboration between professionals with different backgrounds and perspectives. The future professional will be confronted with this transition. Currently, though, degree programmes tend to be highly specific, tailored to individual fields such as landscape

Table 7.1 Key principles of circular economy from a business perspective (based on BSI 2017 and cf. Brad 2018)

CE principle	Explanation
System thinking: understand how your business impacts the whole ecosystem	Companies must consider a holistic approach in product design and manufacturing to understand how individual decisions and activities affect the wider ecosystem, including natural environment, social and economic dimensions.
Innovation: manage resources for more value creation	Companies must innovate in a way that creates business value through the sustainable management of resources incorporated within products and services they design. In other words, this principle strives for connecting economic and environmental gains in product design, manufacturing, and use. It requires business models where companies sell solutions not products, and owning is replaced by sharing.
Stewardship: take responsibility for the ripple-effect impacts that come up from your business activities	Companies have to manage the direct and indirect impacts of their decisions and activities across the systems they create and interact with. Stewardship means a company is responsible for any consequence of its managerial decisions in relation to product design, its production and exploitation, as well as its end-of-life.
Collaboration: secure benefits at system wide level by strong cooperation in the value chain	Companies have to conduct continuous cooperation, both internally and with external stakeholders, through various business arrangements such as to create mutual business value for all stakeholders.
Value optimization: keep materials at the highest value and function quality	Companies have to keep all products, components, and materials at their highest value and utility at all times, such as recirculation to be done with minimal energy consumption. Recirculation, in any form, is not the goal of circular economy. Recirculation is only a mean to create new value in the system from elements that are considered loss or waste. Value added is in cost saving, in lower environmental impact, in higher business resilience, in new revenue streams, and in better relationship with customers.
Transparency: reveal to everyone the environmental impact of all your business activities.	Companies are fully aware and open about decisions and activities that affect their ability to move towards a more sustainable and circular mode of operation and are willing to communicate their effects in a clear, accurate, timely, honest, and complete manner.

architecture, economics, healthcare, social work, urban planning or technology. Our challenge is to prepare these students for their future and to educate them to think across sectoral boundaries and to work transdisciplinary, thereby addressing the grand societal challenges such as climate adaptation, circularity, urban health, citizen participation, and the energy transition. We aim to achieve this by creating an inter-sectoral transdisciplinary minor, called 'Collaboration for the City of the Future', with support of the Comenius Leadership Fellow Programme of the Netherlands Organisation for Scientific Research (NWO).

This Minor creates a learning environment that will serve as a Living Lab (see Sect. 4.10) where we will structurally examine and discover the applicability of transdisciplinary collaboration in education for all programmes at Inholland University and higher professional education as a whole. This Minor will bring together students from various educational backgrounds, such as social work, landscaping, economics, technology, and urban planning. We will teach the students the process skills needed to work in transdisciplinary teams. As a result, they learn to work together across multiple disciplines, united by an over-arching vision for a sustainable, healthy and inclusive city of the future.

Open Access This chapter is licensed under the terms of the Creative Commons Attribution 4.0 International License (http://creativecommons.org/licenses/by/4.0/), which permits use, sharing, adaptation, distribution and reproduction in any medium or format, as long as you give appropriate credit to the original author(s) and the source, provide a link to the Creative Commons license and indicate if changes were made.

The images or other third party material in this chapter are included in the chapter's Creative Commons license, unless indicated otherwise in a credit line to the material. If material is not included in the chapter's Creative Commons license and your intended use is not permitted by statutory regulation or exceeds the permitted use, you will need to obtain permission directly from the copyright holder.

Conclusion 8

This book shows how the most comprehensive societal fault lines of our times are deeply intertwined and confronts us with challenges concerning the security as well as justice of our societies. Increasing wealth inequality, financial crises, ecological crisis, climate change, trade wars, migration issues, and even vulnerabilities to the coronavirus pandemic (related to global dependencies and interconnectedness) can be traced back to two common denominators.

First, the schism between humans and nature and the dominant anthropocentric world view that arose during the Enlightenment era. Second, the capitalist economic logic and in particular the unsustainability of infinite economic growth in a finite world and belief in the infallibility of the free market that arose after the Second World War.

Since the 1970s, many Western countries have too easily subscribed to an economic model that if the market arranges it, then it is better and more efficient. However, this has left us with market-based societies characterized by individualism and self-interest, materialism, privatization, short-termism, and a dogmatic focus on profit and economic growth. The result diminishes social and ecological values and instead prioritizes excessive production, consumption, and depletion of our natural resources and raw materials. This decades-long focus has resulted in loss of biodiversity and key ecosystem functions, as well as environmental degradation, and the depletion of natural resources and raw materials. We now experience first-hand that ecological vulnerability translates into economic and social vulnerability and a complex set of security and justice challenges.

As the scientific evidence mounts, we can conclude with little doubt that humankind has siphoned resources and stressed the ecological balance across planet. We know that loss of key ecosystem functions and biodiversity threatens the well-being of our own species and that effects from global warming and environmental degradation have real consequences for real people and communities in every corner of the world. These ecological crises have struck the very heart of human coexistence and pose serious threats to security and justice for all.

However, this gloomy message need not lead to despair. Human beings still have time to act and are capable of transitioning to more sustainable models of governance and economics. This book proposes new frameworks and approaches, including the concept of Transformative Social-Ecological Innovation (TSEI) and a Natural Social Contract, to help reshape priorities, habits, and decisions for decades to come.

The half-century between 2000 and 2050 will be remembered as a sustainability transition in what has been called the 'Great Mindshift' (Göpel 2016) or the next 'Great Transformation' (Schellnhuber et al. 2011). The changes and innovations refer to a redirection of civilization that recalls the advent of market economies described by Karl Polanyi in *The Great Transformation* (Polanyi 1944). Following the 2008 global credit crisis, the COVID-19 pandemic in 2020–2021 has once again highlighted painful vulnerabilities of today's world. In fact, the coronavirus crisis could become the next major tipping point towards a more sustainable, healthy, and just society.

The outline of a Natural Social Contract serves as a counter-proposal to existing social contracts. A Natural Social Contract implies an existential change in the way humankind lives in and interacts with its social and natural environment. To navigate this transformation, we will have to find new ways to inhabit and cultivate our planet and keep it healthy for future generations.

Generally speaking, a Natural Social Contract reserves a central place for core values such as solidarity, togetherness, collective well-being (as being central to group life), democracy, equity, and social and environmental justice. More specifically, a Natural Social Contract stresses the importance of social and environmental stewardship. After all, everyone is part of a social and natural environment, and the environment is part of each of us. It is noteworthy that values such as stewardship and solidarity have a prominent role in all world religions. For instance, many religions and denominations have various degrees of support for environmental stewardship, which is a theological belief that humans are responsible for taking care of the world, including all life (humans, animals, and nature). Another example comes from New Zealand, where the Maori term Kaitiaki is used for the concept of 'Guardianship', for the sky, the sea, and the land. This concept has been adopted in New Zealand's legislation, allowing Maori communities to be appointed as guardians for a specific area.

The Natural Social Contract overall seeks to promote a new way of thinking designed to mitigate poverty, inequality, social exclusion, and environmental degradation. A tangible vision could serve as a vehicle to identify and create shared and common values during the process of Transformative Social-Ecological Innovation (TSEI). Agreement on these ethical and normative aspects is important for holding actor coalitions together during a transition process and could be achieved through deliberation on shared beliefs and values, shared discourses, common interests, procedural justice, and options for multiple value creation and mutual gains.

Drawing on economic and institutional design lessons from nature, a Natural Social Contract encourages innovative strategies and tests our institutional and economic models against sustainability benchmarks. Design lessons taught by healthy and mature ecosystems deserve special attention, such as those related to complex adaptive systems, adaptive capacity and resilience, resource efficiency, circularity, self-organization, and the networked relationship between all organisms. From an economic perspective, Circular Economy and Regenerative Economy provide examples of economic design based on ecology, where nature shows how circularity is usually organized at the lowest possible level. The latter provides an argument for short and local supply chains, for realizing circularity related to water, food, energy, raw materials, and consumer goods. This means much less dependence on international trade, especially trade that is characterized by long, expensive and environmentally harmful logistics and supply chains, and risks related to market fluctuations and climate change impacts. For businesses it will require a fundamental shift from linear to circular business models. Such a transition would make businesses more climate resilient, since climate change poses wide-ranging threats to business operations, including disruption in production capacity and supply chains, increased operational costs, or the inability to do business. The latter could result in loss of jobs. From an institutional perspective, the governance of a social-ecological system requires new ways of dealing with ambivalence, complexity, uncertainty, and distributed power in societal change, such as adaptive, reflexive, and deliberative approaches to governance.

Examples of institutional design based on ecology include adaptive spatial planning in urban and rural areas, polycentric governance of the commons, and the sustainable co-management of natural resources (e.g. fishing grounds, forests, and agricultural land), urban commons (e.g. social housing, urban gardening or direct farmers-consumers-cooperatives) and cultural resources (e.g. sources of information, knowledge, and culture). This will require discussion, for instance, to decide under what circumstances it is possible to shift from private property to common property and user rights for the joint management of agricultural land or urban spaces. We also know from nature that a one-size-fits-all approach is doomed to fail. Resilience is increased by biological diversity as well as institutional diversity. A Natural Social Contract, therefore, should be tuned to the specific features of local geography, ecology, economies, and cultures.

This book explains how Transformative Social-Ecological Innovation (TSEI) plays a central role in the sustainability transition and humankind's quest for a Natural Social Contract. Transformative Social-Ecological Innovation (TSEI) is defined as 'systemic changes in established patterns of action as well as in structure, including formal and informal institutions and economies, that contribute to sustainability, health, and justice in all social-ecological systems' (definition by author). It is about society aspiring to create a sustainable, regenerative, and healthy future. This will require collective action and effective cooperation between multiple parties, multiple sectors, and multiple levels, as well as institutional change and new modes of governance.

At the core of TSEI lies the engagement and participation of government, businesses, academia, civilians, civil society, media, and the environment, in a process of multi-party deliberation, co-creation, collective learning, and evidence-based decision-making, resembling the quintuple helix innovation model. The quintuple helix shows how democracy and the environment need to be integrated in the wider perspective of the architecture of TSEI and societal transformation more in general.

In a Natural Social Contract, society cannot rely on the market or state alone for solutions to collective problems, nor leave it to individual responsibility. Instead, collective problems need to be resolved with systemic, sustainable, and fair solutions requiring the involvement or strong representation of groups and stakeholders most affected by those problems. Fundamental change must come from within society. In other words, realizing a Natural Social Contract will require a rethink of how society is organized to solve problems at the most appropriate level (the subsidiarity principle) and by new coalitions in horizontal innovation networks.

From an institutional perspective, a societal transformation towards a Natural Social Contract will require new forms of democracy, governance, organization, management, and cooperation. Adaptive, reflexive, and deliberative approaches to governance will be required that focus on addressing ambivalence, complexity, uncertainty, and distributed power in societal change. It will go hand in hand with processes of collective learning in which different parties learn from each other and participate in joint knowledge production, co-creation and systemic co-design in a transdisciplinary approach. Society's capacity to learn is perhaps the most essential property for realizing a societal transformation towards a Natural Social Contract. Proven methods for collective learning and co-creation can help generate mutual trust, develop a shared understanding of problems, resolve conflicts, and find shared solutions that ultimately enable all stakeholders to achieve better results than otherwise attained on their own.

From an economic perspective, the most fundamental systemic change required for realizing a Natural Social Contract is a transition from our current linear economic system (i.e. produce, use and dispose) towards circular and regenerative economies and cultures. The promise of a circular and regenerative economy is to organize circularity, sustainability and social justice at different scales, preferably as an integrated economic and societal task, which involves technological, social, organisational and institutional innovation. In practice this will require a radical change from linear to circular business models characterized by collective and shared value creation. Innovative and hybrid forms of financing, such as revolving energy and sustainability funds, will also support this development. Likewise, the joint management of commons (instead of private ownership) and a sharing economy focused on sharing of access to goods and services could improve efficiency, sustainability, and community values. These would be important systemic changes toward a Natural Social Contract. Furthermore, True Cost Accounting (TCA), by incorporating the hidden social and ecological costs in the price of products and services, will create opportunities to level the playing field between unsustainable, unhealthy, and unfair production and consumption patterns and systems, with more sustainable, healthy, and fair ones. Finally, taxation remains arguably the most effective policy tool for mitigating unsustainable and unhealthy behaviour. Products and services (e.g. carbon tax) along with tax revenues and tax rebates offer positive incentives for sustainable and healthy practices, behaviour, products, and services.

Likewise, taxation is a powerful tool for addressing growing inequality through tax increases on capital and tax decreases on labour.

Based on a literature review I have highlighted key theories and concepts that provide substance to the workings of Transformative Social-Ecological Innovation (TSEI), such as transition studies (4.2), institutional change and the structure-agency debate (3.9), resilience theory and social-ecological systems (3.8), institutional design principles for governing the commons (4.3), design principles from nature (4.4), complex adaptive systems (4.5), adaptive, reflexive, and deliberative approaches to governance, management, and planning (4.6), social learning, policy learning, and transformational learning (4.7), shared value, multiple value creation, and mutual gains approach (4.8), effective cooperation, (4.9), quintuple helix innovation model (3.9), transdisciplinary cooperation, living labs, and citizen science (4.10), and the art of co-creation: approaches, principles, and pitfalls (4.11).

Drawing on the insights from this literature, I argue that studying and advancing Transformative Social-Ecological Innovation (TSEI) should investigate both structure and agency and at decisive moments where both structure and agency intersect (i.e. in series or clusters of closely related action situations). This includes the resulting outputs, outcomes, and impacts. I identify a critical need to focus on the fundamentally political character of Transformative Social-Ecological Innovation and the need for multiple value creation that promotes shared values, mutual gains, and collective well-being among parties in a social-ecological setting.

The TSEI-framework presented in this book helps to diagnose and advance Transformative Social-Ecological Innovation across sectors and disciplines and at different levels of governance. Predecessors of the TSEI-framework have been used successfully in environmental diplomacy and mediation processes in various parts of the world, as well as in advancing transformation processes and institutional change in water management, agriculture, and spatial planning. The TSEI-framework is proposed as an open framework, in the sense that it is open for additional predictors and moderators if they have a documented effect. To this end, it identifies intervention and leverage points and helps formulate sustainable solutions that can include different views as well as changing and competing needs. Overall, the concept of TSEI encourages public officials, business leaders, and the greater public to think more broadly about how society can rethink cooperation to address humankind's greatest challenges.

We now have an opportunity to make better decisions about how to organize our 21st-century society.

The aim of my research group, in collaboration with our partners, is to generate more insights into Transformative Social-Ecological Innovation (TSEI) for a sustainable, healthy, and just society. By doing so, we can together support our common quest for a Natural Social Contract and not simply for the benefit of ourselves but also for our planet and future generations (Fig. 8.1).

Earth, that's us.

Fig. 8.1 "There is no planet B", by climate protesters (Shutterstock)

Open Access This chapter is licensed under the terms of the Creative Commons Attribution 4.0 International License (http://creativecommons.org/licenses/by/4.0/), which permits use, sharing, adaptation, distribution and reproduction in any medium or format, as long as you give appropriate credit to the original author(s) and the source, provide a link to the Creative Commons license and indicate if changes were made.

The images or other third party material in this chapter are included in the chapter's Creative Commons license, unless indicated otherwise in a credit line to the material. If material is not included in the chapter's Creative Commons license and your intended use is not permitted by statutory regulation or exceeds the permitted use, you will need to obtain permission directly from the copyright holder.

Correction to: Conceptual Background of Transformative Social-Ecological Innovation

Correction to:
Chapter 4 in: P. Huntjens, *Towards a Natural Social Contract*,
https://doi.org/10.1007/978-3-030-67130-3_4

Inadvertently, the book was published with an incorrect citation of Fig. 4.5 in page 100. The citation has been now corrected to read as below,

"In particular, the mental models, the bottom layer in Fig. 4.3, determine or maintain the structures and decision-making of a system."

Open Access This chapter is licensed under the terms of the Creative Commons Attribution 4.0 International License (http://creativecommons.org/licenses/by/4.0/), which permits use, sharing, adaptation, distribution and reproduction in any medium or format, as long as you give appropriate credit to the original author(s) and the source, provide a link to the Creative Commons license and indicate if changes were made.

The images or other third party material in this chapter are included in the chapter's Creative Commons license, unless indicated otherwise in a credit line to the material. If material is not included in the chapter's Creative Commons license and your intended use is not permitted by statutory regulation or exceeds the permitted use, you will need to obtain permission directly from the copyright holder.

The updated online version of the chapter can be found at
https://doi.org/10.1007/978-3-030-67130-3_4

© The Author(s) 2021
P. Huntjens, *Towards a Natural Social Contract*,
https://doi.org/10.1007/978-3-030-67130-3_9

References

Aarts, M. N. C. (2018). *Dynamiek en Dependentie in Socio-Ecologische Interacties*. Inaugural speech. Radboud University, 26 oktober 2018.

Adger, W. N., Quinn, T., Lorenzoni, I., Murphy, C., & Sweeney, J. (2013). Changing social contracts in climate-change adaptation. *Nature Climate Change, 3*(4), 330–333.

Adviesraad voor het Wetenschaps- en Technologiebeleid. (2014). De kracht van sociale innovatie. *Advies 84.* ISBN: 9789077005651.

Afshin, A., Sur, P. J., Fay, K. A., Cornaby, L., Ferrara, G., Salama, J. S., et al. (2019). Health effects of dietary risks in 195 countries, 1990–2017: A systematic analysis for the Global Burden of Disease Study 2017. *The Lancet, 393*(10184), 1958–1972.

Agrawal, A. (2002). Common resources and institutional sustainability. In E. Ostrom, T. Dietz, N. Dolšak, P. C. Stern, S. Stovich, & E. U. Weber (Eds.), *The drama of the commons* (pp. 41–86). Washington, DC: National Academy Press.

Agyeman, J., & McLaren, D. (2017). Sharing cities. *Environment: Science and Policy for Sustainable Development, 59*(3), 22–27.

Ahern, J. (2006). *Theories, methods and strategies for sustainable landscape planning*. Dordrecht: Springer.

Ajzen, I. (1991). The theory of planned behavior. *Organizational Behavior and Decision Processes, 50*, 179–211. https://doi.org/10.1016/0749-5978(91)90020-T.

Albert, M. (1992). *Kapitalisme contra kapitalisme*. Amsterdam/Antwerpen: Uitgeverij Contact.

Ali-Khan, F., & Mulvihill, P. R. (2008). Exploring collaborative environmental governance: Perspectives on bridging and actor agency. *Geography Compass, 2*, 1974–1994. https://doi.org/10.1111/j.1749-8198.2008.00179.x.

Alkon, A. H., & Agyeman, J. (Eds.). (2011). *Cultivating food justice: Race, class, and sustainability*. MIT Press.

Allen, P. M., & McGlade, J. M. (1986). Dynamics of discovery and exploitation: The case of the Scotian Shelf groundfish fisheries. *Canadian Journal of Fisheries and Aquatic Sciences, 43*(6), 1187–1200.

Allen, N., & Sharp, T. (2017). Process peace: A new evaluation framework for track II diplomacy. *International Negotiation, 22*(1), 92–122.

Allin, P., & Hand, D. J. (2017). From a system of national accounts to a process of national wellbeing accounting. *International Statistical Review, 85*(2), 355–370.

Anderies, J., Janssen, M., & Ostrom, E. (2004). A framework to analyze the robustness of social-ecological systems from an institutional perspective. *Ecology and Society, 9*(1).

Argyris, C. (1985). *Action science, concepts, methods, and skills for research and intervention*. San Francisco: Jossey-Bass.

Argyris, C. (1999). *On organizational learning* (2nd ed.). Malden, MA: Blackwell Business Publishers.

Argyris, C., & Schön, D. (1974). *Theory in practice. Increasing professional effectiveness*. San Francisco: Jossey-Bass. Landmark statement of 'double-loop' learning' and distinction between espoused theory and theory-in-action.

Argyris, C., & Schön, D. (1978). *Organizational learning: A theory of action perspective*. Reading, MA: Addison Wesley.

Argyris, C., & Schön, D. (1996). *Organizational learning II: Theory, method and practice*. Reading, MA: Addison Wesley.

Arts, B., & Van Tatenhove, J. P. M. (2004). Policy and power: A conceptual framework between the 'old' and 'new' policy idioms. *Policy Sciences, 37*, 339–356. https://doi.org/10.1007/s11077-005-0156-9.

Ashby, J. (2003). Uniting science and participation in the process of innovation – Research for development. In B. Pound, S. Snapp, C. McDougall, & A. Braun (Eds.), *Managing natural resources for sustainable livelihoods – Uniting science and participation*. Earthscan/IDRC.

Aslaksen, I., Bragstad, T., & Ås, B. (2014). Feminist economics as vision for a sustainable future. *Counting on Marilyn Waring: New Advances in Feminist Economics*, 21–36.

Aspenson, A. (2020). *"True" costs for food system reform: An overview of true cost accounting literature and initiatives*. Johns Hopkins Center for a Livable Future.

Avelino, F., & Rotmans, J. (2009). Power in transition: An interdisciplinary framework to study power in relation to structural change. *European Journal of Social Theory, 12*(4), 543569.

Avelino, F., & Wittmayer, J. M. (2016). Shifting power relations in sustainability transitions: A multi-actor perspective. *Journal of Environmental Policy & Planning, 18*(5), 628–649. https://doi.org/10.1080/1523908X.2015.1112259.

Avelino, F., Grin, J., Pel, B., & Jhagroe, S. (2016). The politics of sustainability transitions. *Journal of Environmental Policy Planning, 18*(5), 557–567.

AWVN. (2016). Een nieuw sociaal contract. Vraaggesprek Prof. Dr. Kim Putter, Directeur Sociaal en Cultureel Planbureau. In *Werkgeven*, nummer 4, jaargang 13, winter 2016/2017.

Bache, I., Bartle, I., & Flinders, M. (2016). Multi-level governance. In *Handbook on theories of governance*. Edward Elgar Publishing.

Backus, G. B. C., Meeusen, M. J. G., Dagevos, J. C., van 't Riet, J. P., Bartels, J., Onwezen, M. C., Reinders, M. J., de Winter, M. A., & Grievink, J. W. (2011). *Voedselbalans 2011: Dl. 1 Dynamiek in duurzaam*. Den Haag: LEI, onderdeel van Wageningen UR.

Bagheri, A., & Hjorth, P. (2007). Planning for sustainable development: A paradigm shift towards a process-based approach. *Sustainable Development, 15*(2), 83.

Bain, P. G., Hornsey, M. J., Bongiorno, R., Kashima, Y., & Crimston, D. (2013). Collective futures: How projections about the future of society are related to actions and attitudes supporting social change. *Personality and Social Psychology Bulletin, 39*(4), 523–539.

Bakker, P., Evers, S., Hovens, N., Snelder, H., & Weggeman, M. (2005). Het Rijnlands model als inspiratiebron. *Holland Management Review, 2005*, 72.

Bandura, A. (1977). Self-efficacy: Toward a unifying theory of behavioral change. *Psychological Review, 84*(2), 191.

Barbu, C. M., Florea, D. L., Ogarcă, R. F., & Barbu, M. C. (2018). From ownership to access: How the sharing economy is changing the consumer behavior. *Amfiteatru Economic, 20*(48), 373–387.

Barnett, J. (1997, December 4). *Environmental security: Now what?* Seminar, Department of International Relations, Keele University.

Barnett, J. (2009). Environmental security in the Asia-Pacific region: Contrasting problems, places, and prospects. In *Facing global environmental change* (pp. 939–950). Berlin, Heidelberg: Springer.

Barnett, J., & Adger, W. N. (2007). Climate change, human security and violent conflict. *Political Geography, 26*(6), 639–655.

Barth, T. D. (2011). The idea of a green new deal in a Quintuple Helix Model of knowledge, know-how and innovation. *International Journal of Social Ecology and Sustainable Development (IJSESD), 2*(1), 1–14.

Baum, F., MacDougall, C., & Smith, D. (2006). Participatory action research. *Journal of Epidemiology and Community Health, 60*(10), 854.

Bazerman, M., & Neal, M. A. (1992). The mythical fixed-pie. In *Negotiating rationally* (pp. 16–22). New York, NY: Free Press.

Bazerman, M., & Watkins, M. (2004). Preventing predictable surprises. In *Predictable surprises: The disasters you should have seen coming, and how to prevent them* (pp. 153–258). Boston, MA: Harvard Business School Press.

Becker, C. (2006). The human actor in ecological economics: Philosophical approach and research perspectives. *Ecological Economics, 60*(1), 17–23.

Becker, E. (2012). Social-ecological systems as epistemic objects. In *Human-nature interactions in the Anthropocene* (pp. 55–77). Routledge.

Beers, P., van Mierlo, B., & Hoes, A.-C. (2016). Toward an integrative perspective on social learning in system innovation initiatives. *Ecology and Society, 21*(1), 33.

Beers, P. J., Turner, J. A., Rijswijk, K., Williams, T., Barnard, T., & Beechener, S. (2019). Learning or evaluating? Towards a negotiation-of-meaning approach to learning in transition governance. *Technological Forecasting and Social Change, 145*, 229–239.

Belk, R. (2014). You are what you can access: Sharing and collaborative consumption online. *Journal of Business Research, 67*(8), 1595–1600.

Bennett, C. J., & Howlett, M. (1992). The lessons of learning: Reconciling theories of policy learning and policy change. *Policy Sciences, 25*, 275–294.

Benyus, J. M. (1997). *Biomimicry: Innovation inspired by nature*. HarperCollins.

Berkes, F., & Turner, N. J. (2006). Knowledge, learning and the evolution of conservation practice for social-ecological system resilience. *Human Ecology, 34*(4), 479.

Berkes, F., Folke, C., & Colding, J. (Eds.). (2000). *Linking social and ecological systems: Management practices and social mechanisms for building resilience*. Cambridge University Press.

Berkes, F., Colding, J., & Folke, C. (Eds.). (2002). *Navigating social-ecological systems: Building resilience for complexity and change*. Cambridge, MA: Cambridge University Press.

Berry, T. (1999). *The great work: Our way into the future*. Bell Tower: Crown Publishing Group. ISBN: 978-0609804995.

Berry, T. (2006). *Evening thoughts: Reflecting on Earth as a sacred community*. San Francisco, CA: Sierra Club Books. ISBN: 978-1619025318.

Beunen, R., & Patterson, J. J. (2019). Analysing institutional change in environmental governance: Exploring the concept of 'institutional work'. *Journal of Environmental Planning and Management, 62*(1), 12–29.

Beverland, M. B. (2014). Sustainable eating: Mainstreaming plant-based diets in developed economies. *Journal of Macromarketing, 34*, 369–382.

Biebricher, T. (2017). *Onvermoed en onvermijdelijk – De vele gezichten van het neoliberalisme*. Uitgeverij Valkhof Pers.

Biermann, F., Pattberg, P. H., & Heires, M. (2007). *Governance and institutions for sustainability*. (External report, IVM report, no E-07/07). Amsterdam: Institute for Environmental Studies.

Biggs, S., Lake, O., & Goldtooth, T. (Eds.). (2017). *Rights of nature & mother Earth: Rights-based law for systemic change (PDF) (report)*. Movement Rights, Women's Earth & Climate Action Network, Indigenous Environmental Network. Archived (PDF) from the original on 2018-02-05.

Black, S., Gardner, D. G., Pierce, J. L., & Steers, R. (2019). *Design thinking*. Organizational Behavior.

Boerenbusiness. (2018). Landbouw: hoe ontwikkeld duurzaamheid zich. Retrieved from March 5, 2019, from https://www.boerenbusiness.nl/rss/algemeen/artikel/10880455/landbouw-hoe-ontwikkelt-duur-zaamheid-zich.

Bollier, D., & Helfrich, S. (Eds.). (2015). *Patterns of commoning*. Commons Strategy Group and Off the Common Press.

Bonney, R., Shirk, J. L., Phillips, T. B., Wiggins, A., Ballard, H. L., Miller-Rushing, A. J., & Parrish, J. K. (2014). Next steps for citizen science. *Science, 343*(6178), 1436–1437.

Boonstra, J. J. (2004). Introduction. In J. J. Boonstra (Ed.), *Dynamics of organizational change and learning*. Chichester: Wiley.

Borràs, S. (2016). New transitions from human rights to the environment to the rights of nature. *Transnational Environmental Law, 5*(1), 113–143.

Bos, A. P., & Grin, J. (2012). Reflexive interactive design as an instrument for dual track governance. In *System innovations, knowledge regimes, and design practices towards transitions for sustainable agriculture* (pp. 132–153). INRA.

Bos, A. P., Koerkamp, P. G., Gosselink, J. M. J., & Bokma, S. (2009). Reflexive interactive design and its application in a project on sustainable dairy husbandry systems. *Outlook on Agriculture, 38*(2), 137–145.

Bosselmann, K. (2004). In search of global law: The significance of the Earth Charter. *Worldviews: Global Religions, Culture, and Ecology, 8*(1), 62–75.

Boulanger, P., & Philippidis, G. (2015). The EU budget battle: Assessing the trade and welfare impacts of CAP budgetary reform. *Food Policy, 51*, 119–130.

Bouma, J., Koetse, M., & Brandsma, J. (2020). *Natuurinclusieve landbouw: wat beweegt boeren. Het effect van financiële prikkels en gedragsfactoren op de investeringsbereidheid van agrariërs*. Den Haag: Planbureau voor de Leefomgeving.

Bouman, T., & Steg, L. (2019). Motivating society-wide pro-environmental change. *One Earth, 1*(1), 27–30.

Bourdieu, P. (1988). Vive la crise! *Theory and Society, 17*(5), 773–787.

Bourdieu, P. (2005). *The social structures of the economy*. Polity.

Brad, S. (2018, June). *Circular economy innovation tools – Principles of circular economy*. Qualification Programme Handbook, MOVECO.

Brauch, H. G., & Scheffran, J. (2012). Introduction: Climate change, human security, and violent conflict in the Anthropocene. In *Climate change, human security and violent conflict* (pp. 3–40). Berlin, Heidelberg: Springer.

Bremmer, B., & Bos, B. (2017). *Creating niches by applying reflexive interactive design. Agro-Ecological transitions*. Wageningen: Wageningen University & Research.

Brown, J. S. (2000). Growing up digital: How the Web changes work, education, and the ways people learn. *Changes, 32*(2), 10–20.

Brown, T. (2008). Design thinking. *Harvard Business Review, 86*(6), 84.

Brown, V., Moodie, M., Cobiac, L., Herrera, A. M., & Carter, R. (2017). Obesity-related health impacts of fuel excise taxation – An evidence review and cost-effectiveness study. *BMC Public Health, 17*(1), 359.

Brugnach, M., Dewulf, A., Pahl-Wostl, C., & Taillieu, T. (2008). Toward a relational concept of uncertainty: About knowing too little, knowing too differently, and accepting not to know. *Ecology and Society, 13*(2), 30. [online].

Bryson, J. (1988). Strategic planning: Big wins and small wins. *Public Money and Management, 8*(3), 11–15.

Brzoska, M. (2012). Climate change as a driver of security policy. In *Climate change, human security and violent conflict* (1st ed., pp. 165–184). Heidelberg: Springer.

BSI. (2017). *British Standards Institution BS 8001: 2017 Framework for implementing the principles of the circular economy in organizations – Guide*. The British Standards Institution.

Burgerboerderijen. (2019). *Burgerboerderijen*. Retrieved from March 14, 2019, from https://burgerboerderijen.nl/.

C40 Cities. (2018, December). *Municipality-led circular economy case studies. In partnership with the Climate-KIC Circular Cities Project*. C40 Cities.

Calhoun, C. (2002). *Dictionary of the social sciences*. Oxford: Oxford University Press.

Campbell, B. M., Vermeulen, S. J., Aggarwal, P. K., Corner-Dolloff, C., Girvetz, E., Loboguerrero, A. M., et al. (2016). Reducing risks to food security from climate change. *Global Food Security, 11*, 34–43.

Caradonna, J., Borowy, I., Green, T., Victor, P. A., Cohen, M., Gow, A., & Heinberg, R. (2015). A degrowth response to an ecomodernist manifesto. *Resilience.org*.

Carayannis, E. G., & Campbell, D. F. (2010). Triple Helix, Quadruple Helix and Quintuple Helix and how do knowledge, innovation and the environment relate to each other?: A proposed framework for a trans-disciplinary analysis of sustainable development and social ecology. *International Journal of Social Ecology and Sustainable Development (IJSESD), 1*(1), 41–69.

Carpenter, S., Walker, B., Anderies, J. M., & Abel, N. (2001). From metaphor to measurement: Resilience of what to what? *Ecosystems, 4*(8), 765–781.

Cattacin, S., & Zimmer, A. (2016). Urban governance and social innovations. In T. Brandsen, S. Cattacin, A. Evers, & A. Zimmer (Eds.), *Social innovations in the urban context* (pp. 21–44). Heidelberg: Springer.

CBS. (2018). *Monitor Brede Welvaart 2018*. Den Haag/Heerlen/Bonaire: Centraal Bureau voor de Statistiek. ISBN: 978-90-357-2128-9.

CBS. (2020). *Monitor Brede Welvaart & sustainable development goals 2020*. Den Haag/Heerlen/Bonaire: Centraal Bureau voor de Statistiek.

CDP. (2017). *The Carbon Majors Database*. CDP Carbon Majors Report.

Cecchi, C. (2013). Sostenibilità e Decrescita: dall'Homo Œconomicus all'Homo Ecologicus (Sustainability and de-growth: From the homo economicus to the homo ecologicus). In E. Basile, G. Lunghini, & Volpi, F. (A Cura Di) (Eds.), *Pensare il Capitalismo. Nuove Prospettive per l'Economia Politica (Thinking about capitalism. New perspectives for political economy)* (pp. 168–184). Milano: Francoangeli.

Chaffin, B. C., Gosnell, H., & Cosens, B. A. (2014). A decade of adaptive governance scholarship: Synthesis and future directions. *Ecology and Society, 19*(3).

Chambers, S. (2018). Human life is group life: Deliberative democracy for realists. *Critical Review, 30*(1–2), 36–48.

Chandler, M., See, L., Copas, K., Bonde, A. M., López, B. C., Danielsen, F., et al. (2017). Contribution of citizen science towards international biodiversity monitoring. *Biological Conservation, 213*, 280–294.

Change Magazine. (2009). *Samen blijven leren in het waterbeheer*. Retrieved from http://www.changemagazine.nl/klimaatkennis/water/samen_blijven_leren_in_het_waterbeheer.

Checkland, P., & Holwell, S. (1998). Action research: Its nature and validity. *Systemic Practice and Action Research, 11*(1).

Chesbrough, H. W. (2006). Open innovation: A new paradigm for understanding industrial. In H. W. Chesbrough, W. Vanhaverbeke, & J. West (Eds.), *Open innovation: Researching a new paradigm*. Oxford University Press.

Chomsky, N. (1999 [1970]). Language and freedom. *Resonance, 4*, 85–104.

Cialdini, R. B., Reno, R. R., & Kallgren, C. A. (1990). A focus theory of normative conduct: Recycling the concept of norms to reduce littering in public places. *Journal of Personality and Social Psychology, 58*(6), 1015.

Cilliers, P. (2000). What can we learn from a theory of complexity? *Emergence, 2*(1), 23–33. https://doi.org/10.1207/S15327000EM0201_03.

Clark, K. H., & Nicholas, K. A. (2013). Introducing urban food forestry: A multifunctional approach to increase food security and provide ecosystem services. *Landscape Ecology, 28*(9), 1649–1669.

Clark, W. C., Tomich, T. P., Van Noordwijk, M., Guston, D., Catacutan, D., Dickson, N. M., & McNie, E. (2016). Boundary work for sustainable development: Natural resource management at the Consultative Group on International Agricultural Research (CGIAR). *Proceedings of the National Academy of Sciences, 113*(17), 4615–4622.

Clarysse, B., Wright, M., Bruneel, J., & Mahajan, A. (2014). Creating value in ecosystems: Crossing the chasm between knowledge and business ecosystems. *Research Policy, 43*(7), 1164–1176.

CNA Military Advisory Board. (2007). *National security and the threat of climate change*. Alexandria, VA: CNA Corporation. Retrieved from http://www.cna.org/sites/default/files/National%20Security%20and%20the%20Threat%20of%20Climate%20Change%20-%20Print.pdf.

CoE. (2008). *12 principles of good democratic governance*. Council of Europe (CoE). Retrieved March 16, 2020, from https://rm.coe.int/12-principles-brochure-final/1680741931.
Coghlan, D., & Brannick, T. (2002). *Doing action research in your own organization*. London: Sage Publications. Reason & Bradbury, 2010.
Colding, J., Barthel, S., Bendt, P., Snep, R., van der Knaap, W., & Ernstson, H. (2013). Urban green commons: Insights on urban common property systems. *Global Environmental Change, 23*(5), 1039–1051.
Compston, H., & Madsen, P. K. (2001). Conceptual innovation and public policy: Unemployment and paid leave schemes in Denmark. *Journal of European Social Policy, 11*(2), 117–132.
Consensus Building Insitute (CBI). (2014). CBI's mutual gains approach to negotiation. *CBI-MGA Brief*.
Cook, T. D., & Campbell, D. T. (1979). *Quasi-experimentation. Design and analysis issues for field settings*. Boston: Houghton Mifflin.
Costanza, R. (1989). What is ecological economics? *Ecological Economics, 1*, 1–7. https://doi.org/10.1016/0921-8009(89)90020-7.
Cox, M., Arnold, G., & Villamayor Tomás, S. (2010). A review of design principles for community-based natural resource management. *Ecology and Society, 15*(4), 38. [online].
Craps, M. (Ed.). (2003). *Social learning in river basin management*. Report of workpackage 2 of the HarmoniCOP project. Retrieved from www.harmonicop.info.
Creighton, J. L., Priscoli, J. D., Mark Dunning, C., & Ayres, D. B. (1998). *Public involvement and dispute resolution –Volume 2: A reader on the second decade of experience at the institute for water resources*. Alexandria, VA: Institute for Water Resources. U.S. Army Corps of Engineers.
Cullinan, C. (2011). *Wild law: A manifesto for Earth justice* (2nd ed.). USA: Chelsea Green Publishing. ISBN: 978-1603583770.
Cuperus, F., Smit, E., Faber, J., Casu, F., Schütt, J., van Rooij, S., et al. (2019). *Verkenning kennisbehoeftes van agrariërs tav natuurinclusieve landbouw en het reeds bestaande aanbod van deze kennis: waar is de match, de mismatch en hoe die te overbruggen* (No. WPR-797). Stichting Wageningen Research, Wageningen Plant Research, Business Unit Open Teelten.
Dabelko, G. D., & Conca, K. (Eds.). (2019). *Green planet blues: Critical perspectives on global environmental politics*. Routledge.
Dagevos, H. (2018). Sociale innovatie: sociologische traditie en voedseltransitie. In P. M. Karré, H. Dagevos, & G. Walraven (Eds.), *Sociale innovatie in de praktijk: Zoeken naar nieuwe antwoorden op maatschappelijke vraagstukken*. Koninklijke Van Gorcum.
Daly, H. E. (1973). *Toward a steady-state economy* (Vol. 2). San Francisco: WH Freeman.
Dawson, T. P., Perryman, A. H., & Osborne, T. M. (2016). Modelling impacts of climate change on global food security. *Climatic Change, 134*(3), 429–440.
De Leeuw, A., Valois, P., Ajzen, I., & Schmidt, P. (2015). Using the theory of planned behavior to identify key beliefs underlying pro-environmental behavior in high-school students: Implications for educational interventions. *Journal of Environmental Psychology, 42*, 128–138.
De Witte, M., & Jonker, J. (2006). *Management models for corporate social responsibility*. Heidelberg: Springer.
Degenaar, J. (2019). *Barriers in the transition towards nature inclusive agriculture*. Doctoral dissertation.
Delaney, D., & Leitner, H. (1997). The political construction of scale. *Political Geography, 16*(2), 93–97.
Demaria, F., Schneider, F., Sekulova, F., & Martinez-Alier, J. (2013). What is degrowth? From an activist slogan to a social movement. *Environmental Values, 22*, 191–215. https://doi.org/10.2307/23460978.
Desing, H., Brunner, D., Takacs, F., Nahrath, S., Frankenberger, K., & Hischier, R. (2020). A circular economy within the planetary boundaries: Towards a resource-based, systemic approach. *Resources, Conservation and Recycling, 155*, 104673.

Di Gregorio, M., Fatorelli, L., Paavola, J., Locatelli, B., Pramova, E., Nurrochmat, D. R., et al. (2019). Multi-level governance and power in climate change policy networks. *Global Environmental Change, 54*, 64–77.

DiMaggio, P. J., & Powell, W. W. (1983). The iron cage revisited: Institutional isomorphism and collective rationality in organizational fields. *American Sociological Review, 48*, 147–160.

DiMaggio, P. J., & Powell, W. W. (2000). The iron cage revisited institutional isomorphism and collective rationality in organizational fields. *Economics Meets Sociology in Strategic Management, 17*, 143–166.

Distrifood Dynamics. (2019). *Online database met informatie over supermarkten in Nederland.* Retrieved January 24, 2019, from https://distrifooddynamics.nl.

Dobson, A., & Eckersley, R. (Eds.). (2006). *Political theory and the ecological challenge.* Cambridge University Press.

Dorst, K. (2011). The core of 'design thinking' and its application. *Design Studies, 32*(6), 521–532.

Doyle, C., David, R., Li, Y., Luczak-Roesch, M., Anderson, D., & Pierson, C. M. (2019). Using the web for science in the classroom: Online citizen science participation in teaching and learning. In *Proceedings of the 10th ACM conference on web science* (pp. 71–80).

Dryzek, J. S. (1996). Foundations for environmental political economy: The search for homo ecologicus? *New Political Economy, 1*(1), 27–40.

Dryzek, J. S., Hunold, C., Schlosberg, D., Downes, D., & Hernes, H. K. (2002). Environmental transformation of the state: The USA, Norway, Germany and the UK. *Political Studies, 50*(4), 659–682.

Dryzek, J. S., Bächtiger, A., Chambers, S., Cohen, J., Druckman, J. N., Felicetti, A., et al. (2019). The crisis of democracy and the science of deliberation. *Science, 363*(6432), 1144–1146.

Dunleavy, P., Margetts, H., Bastow, S., & Tinkler, J. (2006, July 1). New public management is dead—Long live digital-era governance. *Journal of Public Administration Research and Theory, 16*(3), 467–494. https://doi.org/10.1093/jopart/mui057.

Dupuy, P. M., & Viñuales, J. E. (2018). *International environmental law.* Cambridge University Press.

EAT Lancet. (2018). *The healthy diets from sustainable food systems.* https://eatforum.org/content/uploads/2019/07/EATLancet_Commission_Summary_Report.pdf.

Ecopolis. (2020). *Heating the sustainable city.* Retrieved October 9, 2020, from http://ecopolis.danfoss.com/.

Eden, C., & Huxham, C. (1996). Action research for the study of organizations (Chapter 3.2). In S. R. Clegg, C. Hardy, & W. R. Nord (Eds.), *Handbook of organization studies* (pp. 526–542). London: Sage.

EEA-PZH. (2017). *Transitieaanpak toegepast ten behoeve van een duurzame landbouw in de Provincie Zuid-Holland.* Interne studie van de Eenheid Audit en Advies (EEA) van de Provincie Zuid-Holland.

Ekins, P. (2000). *Economic growth and environmental sustainability: The prospects for green growth.* Psychology Press.

Ellen MacArthur Foundation. (2015). *Towards a circular economy: Business rationale for an accelerated transition.*

Ellen MacArthur Foundation. (2019). *Cities and circular economy for food.*

Elmqvist, T., Gatzweiler, F., Lindgren, E., & Liu, J. (2019). Resilience management for healthy cities in a changing climate. In *Biodiversity and health in the face of climate change* (pp. 411–424). Cham: Springer.

Eriksen, S. H., Nightingale, A. J., & Eakin, H. (2015). Reframing adaptation: The political nature of climate change adaptation. *Global Environmental Change, 35*(523), 533.

Erisman, J. W., van Eekeren, N., Cuijpers, W., & de Wit, J. (2014). *Biodiversiteit in de melkveehouderij. Investeren in veerkracht en reduceren van risico's.* Driebergen: Louis Bolk Instituut.

Erisman, J. W., van Eekeren, N., de Wit, J., Koopmans, C., Cuijpers, W., Oerlemans, N., & Koks, B. J. (2016). Agriculture and biodiversity: A better balance benefits both. *AIMS Agriculture and Food, 1*(2), 157–174.

Erisman, J. W., van Eekeren, N., van Doorn, A., Geertsema, W., & Polman, N. (2017). *Maatregelen Natuurinclusieve landbouw*. (Wageningen Environmental Research rapport; No. 2821). Wageningen: Wageningen Environmental Research.

van Etten, J., de Sousa, K., Aguilar, A., Barrios, M., Coto, A., Dell'Acqua, M., Fadda, C., Gebrehawaryat, Y., van de Gevel, J., Gupta, A., Kiros, A. Y., Madriz, B., Mathur, P., Mengistu, D. K., Mercado, L., Mohammed, J. N., Paliwal, A., Pè, M. E., Quirós, C. F., Rosas, J. C., Sharma, N., Singh, S. S., Solanki, I. S., & Steinke, J. (2019). Crop variety management for climate adaptation supported by citizen science. *Proceedings of the National Academy of Sciences, 116*(10), 4194–4199.

Etzkowitz, H., & Leydesdorff, L. (2000). The dynamics of innovation: From National Systems and "Mode 2" to a Triple Helix of university–industry–government relations. *Research Policy, 29*(2), 109–123.

EU, 7th Environment Action Programme. (2013). *Decision No 1386/2013/EU of the European Parliament and of the Council of 20 November 2013 on a General Union Environment Action Programme to 2020 'Living well, within the limits of our planet'*. Brussel.

EU Commission. (2014). *Towards a circular economy: A zero waste programme for Europe*. Brussels.

European Commission. (2011). *Communication from the commission to the European Parliament, the Council, the European Economic and Social Committee and the Committee of the Regions. A roadmap for moving to a competitive low carbon economy in 2050*. COM (2011) 112/4. Brussels, Belgium.

Ezzati, M., Vander Hoorn, S., Lawes, C. M., Leach, R., James, W. P. T., Lopez, A. D., et al. (2005). Rethinking the "diseases of affluence" paradigm: Global patterns of nutritional risks in relation to economic development. *PLoS Medicine, 2*(5), e133.

FAO, IFAD, UNICEF, WFP, & WHO. (2019). *The state of food security and nutrition in the world 2019. Safeguarding against economic slowdowns and downturns*. Rome: FAO.

Fatheuer, T., Fuhr, L., & Unmüßig, B. (2015). *Kritik der grünen Ökonomie*. oekom Verlag.

Feindt, P. H., & Weiland, S. (2018). Reflexive governance: Exploring the concept and assessing its critical potential for sustainable development. Introduction to the special issue. *Journal of Environmental Policy & Planning, 20*(6), 661–674.

Festinger, L. (1954). A theory of social comparison processes. *Human Relations, 7*(2), 117–140.

Fios, F., & Arivia, G. (2018). The concept of homo ecologicus spiritual-ethical (An ethical reflection on the ecological humanism concept of Henryk Skolimowski). In *Cultural dynamics in a globalized world*. Routledge.

Fisher, R., Ury, W., & Patton, B. (1991). What if they are more powerful? (Developing your BATNA – Best alternative to negotiated agreement). In *Getting to YES: Negotiating agreement without giving in* (2nd ed., pp. 97–107). New York, NY: Penguin Books USA Inc.

Fladerer, J. P., & Kurzmann, E. (2019). *The wisdom of the many: How to create self-organisation and how to use collective intelligence in companies and in society from management to managemANT*. BoD–Books on Demand.

Fleurbaey, M., & Ponthière, G. (2019). *Measuring well-being and lives worth living*. EconPapers.

Foley, D. K. (1998). Introduction (Chapter 1). In P. S. Albin (Ed.), *Barriers and bounds to rationality: Essays on economic complexity and dynamics in interactive systems*. Princeton: Princeton University Press.

Folke, C., Hahn, T., Olsson, P., & Norberg, J. (2005). Adaptive governance of social-ecological systems. *Annual Review of Environment and Resources, 30*, 8.1–8.33.

Food and Agriculture Organisation of the United Nations (FAO). (2016). *Climate change and food security: Risks and responses*. Rome: FAO. ISBN: 978-92-5-108998-9.

Food Security Information Network (FSIN). (2020). *2020 Global report on food crises: Joint analysis for better decisions*. Rome, Italy and Washington, DC: Food and Agriculture Organization (FAO); World Food Programme (WFP); and International Food Policy Research Institute (IFPRI). Retrieved from https://www.fsinplatform.org/global-report-food-crises-2020.

Foster, S., & Iaione, C. (2018). Ostrom in the city: Design principles and practices for the urban commons. In D. Cole, B. Hudson, & J. Rosenbloom (Eds.), *Routledge handbook of the study of the commons*. Routledge.

Frantzeskaki, N., Dumitru, A., Anguelovski, I., Avelino, F., Bach, M., Best, B., et al. (2016). Elucidating the changing roles of civil society in urban sustainability transitions. *Current Opinion in Environmental Sustainability, 22*, 41–50.

Fratini, C. F., Georg, S., & Jørgensen, M. S. (2019). Exploring circular economy imaginaries in European cities: A research agenda for the governance of urban sustainability transitions. *Journal of Cleaner Production, 228*, 974–989.

Fredrickson, B. L., & Losada, M. F. (2005). Positive affect and the complex dynamics of human flourishing. *American psychologist, 60*(7), 678.

Freudenstein, J. V., Broe, M. B., Folk, R. A., & Sinn, B. T. (2017). Biodiversity and the species concept—lineages are not enough. *Systematic Biology, 66*(4), 644–656.

Fukuyama, F. (1992). *The end of history and the last man*. Free Press. ISBN: 0-02-910975-2.

Fullerton, J. (2015). *Regenerative capitalism*. Greenwich, CT: Capital Institute.

Gabriels, R. (2018). John Rawls. In Doorman & Pott (Eds.), *Filosofen van deze tijd* (1st ed.). Uitgeverij Prometheus, 544 p. ISBN: 9789044637373.

Galbraith, J. K. (2012). *Inequality and instability: A study of the world economy just before the great crisis*. Oxford University Press.

Gallopín, G. C. (2006). Linkages between vulnerability, resilience, and adaptive capacity. *Global Environmental Change, 16*(3), 293–303.

Galt, R. E. (2013). The moral economy is a double-edged sword: Explaining farmers' earnings and self-exploitation in community-supported agriculture. *Economic Geography, 89*(4), 341–365.

Gardner, G. T., Prugh, T., & Renner, M. (2016). *Can a city be sustainable?* Island Press.

Gassmann, O., Enkel, E., & Chesbrough, H. (2010). The future of open innovation. *R&D Management, 40*(3), 213–221.

Gauger, A., Rabatel-Fernel, M. P., Kulbicki, L., Short, D., & Higgins, P. (2012). *Ecocide is the missing 5th crime against peace*. London: Human Rights Consortium, School of Advanced Study, University of London. Retrieved from https://sas-space.sas.ac.uk/4830/1/Ecocide_research_report_19_July_13.pdf.

Geels, F. W. (2011). The multi-level perspective on sustainability transitions: Responses to seven criticisms. *Environmental Innovation and Societal Transitions, 1*(1), 24–40.

Geels, F. W., & Kemp, R. (2000). *Transities vanuit sociotechnisch perspectief*. Report for the Dutch Ministry of Environment Universiteit Twente, and Maastricht: MERIT. Maastricht.

Geels, F. W., & Schot, J. (2007). Typology of sociotechnical transition pathways. *Research Policy, 36*(3), 399417.

Geels, F., Elzen, B., & Green, K. (2004). General introduction: Systems innovation and transitions to sustainability. In B. Elzen, F. Geels, & K. Green (Eds.), *Systems innovation and the transition to sustainability: Theory, evidence and policy* (pp. 1–16). Cheltenham: Edward Elgar.

Gerbens-Leenes, P. W., Mekonnen, M. M., & Hoekstra, A. Y. (2013). The water footprint of poultry, pork and beef: A comparative study in different countries and production systems. *Water Resources and Industry, 1–2*, 25–36. https://doi.org/10.1016/j.wri.2013.03.001.

van Geuns, L., Slingerland, S., Bolscher, H., & de Jong, S. (2016, March). *Het fossiele dilemma van Rotterdam*. TNO discussiepaper.

Giddens, A. (1984). *The constitution of society. Outline of the theory of structuration*. Cambridge: Polity Press.

Gies, E., Kros, H., & Voogd, J. C. (2019). *Inzichten stikstofdepositie op natuur*. Wageningen Environmental Research.

Gilbertson, T., & Reyes, O. (2009, November). Carbon trading – How it works and why it fails. *Critical Currents, 7*. ISBN/ISSN: 1654-4250.

Gill, S., Handley, J., Ennos, R., & Pauleit, S. (2007). Adapting cities for climate change: The role of the green infrastructure. *Journal of the Built Environment, 33*(1), 115–133.

Gilmore, T., Krantz, J., & Ramirez, R. (1986, Fall). Action based modes of inquiry and the host-researcher relationship. *Consultation, 5*, 3.

Girardet, H. (2017). Regenerative cities. In *Green economy reader* (pp. 183–204). Cham: Springer.

Glenn, J. C., Gordon, T. J., & Perelet, R. (1998). *Defining environmental security: Implications for the US army*. Atlanta, GA: Army Environmental Policy Inst.
Glopan. (2016). Global panel on agriculture and food systems for nutrition. In *Food systems and diets: Facing the challenges of the 21st century*. London, UK. http://glopan.org/sites/default/files/ForesightReport.pdf.
Glover, J. D., Reganold, J. P., Bell, L. W., Borevitz, J., Brummer, E. C., Buckler, E. S., Cox, C. M., Cox, T. S., Crews, T. E., Culman, S. W., DeHaan, L. R., Eriksson, D., Gill, B. S., Holland, J., Hu, F., Hulke, B. S., Ibrahim, A. M. H., Jackson, W., Jones, S. S., Murray, S. C., Paterson, A. H., Ploschuk, E., Sacks, E. J., Snapp, S., Tao, D., Van Tassel, D. L., Wade, L. J., Wyse, D. L., & Xu, Y. (2010). Increased food and ecosystem security via perennial grains. *Science, 328*(5986), 1638–1639.
Godfray, H. C. J., Beddington, J. R., Crute, I. R., Haddad, L., Lawrence, D., Muir, J. F., et al. (2010). Food security: The challenge of feeding 9 billion people. *Science, 327*(5967), 812–818.
Goldstein, B., Birkved, M., Fernandez, J., & Hauschild, M. (2017). Surveying the environmental footprint of urban food consumption. *Journal of Industrial Ecology, 21*(1), 151–165.
Goodijk, R. (2009). *Herwaardering van de Rijnlandse principes: over governance, overleg en engagement*. Uitgever: Gorcum B.V., Koninklijke van. ISBN: 9789023244455.
Göpel, M. (2016). *The great mindshift: How a new economic paradigm and sustainability transformations go hand in hand* (Vol. 2). Springer.
Gorissen, L., Spira, F., Meynaerts, E., Valkering, P., & Frantzeskaki, N. (2018). Moving towards systemic change? Investigating acceleration dynamics of urban sustainability transitions in the Belgian City of Genk. *Journal of Cleaner Production, 173*, 171–185.
Grau, H. R., & Aide, M. (2008). Globalization and land-use transitions in Latin America. *Ecology and Society, 13*(2).
Green, D., & Shapiro, I. (1996). *Pathologies of rational choice theory: A critique of applications in political science*. Yale University Press.
Grey, D., Sadoff, C., & Connors, G. (2009). Effective cooperation on transboundary waters: A practical perspective. In A. Jägerskog & M. Zeitoun (Eds.), *Getting transboundary water right: Theory and practice for effective cooperation*. Stockholm: SIWI.
Grey, D., Andersen, I., Abrams, L., Alam, U., Barnett, T., Kjellén, B., McCaffrey, S., Sadoff, C., Whittington, S. D., & Wolf, A. T. (2010). Sharing water, sharing benefits: Working towards effective transboundary water resources management. In A. T. Wolf (Ed.), *A graduate/professional skills-building workbook*. UNESCO. Oregon State University.
Grin, J. (2010). Understanding transitions from a governance perspective, part III (pp. 223–319) In J. Grin, J. Rotmans, & J. Schot, (in collaboration with F. Geels and D. Loorbach) (Eds.), Transitions to sustainable development – New directions in the study of long term transformative change. New York: Routledge.
Grin, J. (2011). The politics of transition: Conceptual understanding and implications for transition management. *International Journal of Sustainable Development, 14*.
Grin, J. (2016). Transition studies: Basic ideas and analytical approaches. In *Handbook on sustainability transition and sustainable peace*. Switzerland: Sprinter International Publishing.
Grin, J., & Loeber, A. (2007). Theories of policy learning: Agency, structure and change (Chapter 15). In F. Fischer, G. J. Miller, & M. S. Sidney (Eds.), *Handbook of public policy analysis. Theory, politics, and methods* (pp. 201–219). CRC Press–Taylor & Francis Group.
Grin, J., Felix, F., Bos, B., & Spoelstra, S. (2004). Practices for reflexive design: Lessons from a Dutch programme on sustainable agriculture. *International Journal of Foresight and Innovation Policy, 1*(1–2), 146–169.
Grin, J., Rotmans, J., & Schot, J. (2010). *Transitions to sustainable development: New directions in the study of long term transformative change*. Routledge.
GroenPact. (2016). Ontwikkelagenda groen onderwijs 2016–2025: Investeren in een vitaal onderwijssysteem voor een krachtige groene sector. *GroenPact*.

Grunert, K. G., Sonntag, W. I., Glanz-Chanos, V., & Forum, S. (2018). Consumer interest in environmental impact, safety, health and animal welfare aspects of modern pig production: Results of a cross-national choice experiment. *Meat Science, 137*, 123–129.
Gunderson, L. H., & Holling, C. S. (Eds.). (2002). *Panarchy: Understanding transformations in human and natural systems*. Washington, DC: Island Press.
Gunderson, L. H., Holling, C. S., & Light, S. S. (Eds.). (1995). *Barriers and bridges to the renewal of ecosystems and institutions*. New York: Columbia University Press.
Gupta, C., Campbell, D., Munden-Dixon, K., Sowerwine, J., Capps, S., Feenstra, G., & Kim, J. V. S. (2018). Food policy councils and local governments: Creating effective collaboration for food systems change. *Journal of Agriculture, Food Systems, and Community Development, 8*(B), 11–28.
Haberl, H., Fischer-Kowalski, M., Krausmann, F., Martinez-Alier, J., & Winiwarter, V. (2011). A socio-metabolic transition towards sustainability? Challenges for another great transformation. *Sustainable Development, 19*(1), 1–14.
Hajer, M., & Poorter, M. (2005, July). *Visievorming in transitieprocessen. Een evaluatieonderzoek in opdracht van het Milieu- en Natuurplanbureau/RIVM*. Universiteit van Amsterdam – ASSR.
Hall, P. A. (1988). *Policy paradigms, social learning and the state*. A paper presented to the International Political Science Association, Washington, DC.
Hallegatte, S., Heal, G., Fay, M., & Treguer, D. (2011). *From growth to green growth-a framework*. The World Bank.
Hand, E. (2010). Citizen science: People power. *Nature, 466*(7307), 685–687. https://doi.org/10.1038/466685a.
Hanks, C. A. (2006). Community empowerment: A partnership approach to public health program implementation. *Policy, Politics, & Nursing Practice, 7*(4), 297–306.
Hargrove, R. (2002). *Masterful coaching* (Revised ed.). San Francisco: Jossey-Bass Pfeiffe.
Havelaar, A. H., Kirk, M. D., Torgerson, P. R., Gibb, H. J., Hald, T., Lake, R. J., et al. (2015). World Health Organization global estimates and regional comparisons of the burden of foodborne disease in 2010. *PLoS Medicine, 12*(12), e1001923.
Haxeltine, A., Avelino, F., Pel, B., Dumitru, A., Kemp, R., Longhurst, N., Chilvers, J., & Wittmayer, J. M. (2016). *A framework for Transformative Social Innovation* (TRANSIT Working Paper # 5). TRANSIT: EU SSH.2013.3.2-1 Grant agreement no: 613169.
Hekkert, M., & Ossebaard, M. (2010, March). *De innovatiemotor: Het versnellen van baanbrekende innovaties* (2nd ed.). Gorcum B.V., Koninklijke van. Paperback, 122 p.
Heldt, S., Rodríguez-de-Francisco, J. C., Dombrowsky, I., Feld, C. K., & Karthe, D. (2017). Is the EU WFD suitable to support IWRM planning in non-European countries? Lessons learnt from the introduction of IWRM and River Basin Management in Mongolia. *Environmental Science & Policy, 75*, 28–37.
Hendriks, C. M., & Grin, J. (2007). Contextualizing reflexive governanance: The politics of Dutch transitions to sustainability. *Journal of Environmental Policy Planning, 9*(34), 333–350.
Heo, J., & Muralidharan, S. (2019). What triggers young Millennials to purchase eco-friendly products?: The interrelationships among knowledge, perceived consumer effectiveness, and environmental concern. *Journal of Marketing Communications, 25*(4), 421–437.
Herenboeren. (2019). *Herenboeren*. Retrieved October 12, 2020, from https://www.herenboeren.nl.
Hickey, S. (2011). The politics of social protection: What do we get from a 'social contract'approach? *Canadian Journal of Development Studies/Revue canadienne d'études du développement, 32*(4), 426–438.
Higgins, P. (2010). *Eradicating ecocide*. London: Shepheard-Walwyn Publishers.
Higgins, P., Short, D., & South, N. (2013). Protecting the planet: A proposal for a law of ecocide. *Crime, Law and Social Change, 59*(3), 251–266.
Hisschemöller, M. (2005). Participation as knowledge production and the limits of democracy. In S. Maasen & P. Weingart (Eds.), *Democratisation of expertise? Exploring novel forms of scientific advice in political decision-making* (pp. 189–208). Dordrecht: Springer.

Hoekstra, R. (2019). *Replacing GDP by 2030: Towards a common language for the well-being and sustainability community.* Cambridge University Press.

Hoffmann, U. (2015). *Can green growth really work and what are the true (socio-) economics of Climate Change? United Nations Conference on Trade and Development (UNCTAD).* CH-Geneva.

Hogarth, P. J. (2015). *The biology of mangroves and seagrasses.* Oxford University Press.

Hogwood, B. W., & Peters, B. G. (1983). *Policy dynamics.* Brighton: Wheatsheaf Books.

Holland, J. H. (1995). *Hidden order – How adaptation builds complexity* (No. 003.7 H6).

Holland, J. H. (2006). Studying complex adaptive systems. *Journal of Systems Science and Complexity, 19*(1), 1–8.

Holling, C. S. (1978). *Adaptive environmental assessment and management.* London: John Wiley.

Hooghe, L., & Marks, G. (2003, May). Unraveling the Central State, but how? Types of multi-level governance. *American Political Science Review, 97*(2).

Hooghe, L., Marks, G., & Schakel, A. (2020). Multilevel governance. *Comparative Politics, 193.*

Hopkins, F. M., Ehleringer, J. R., Bush, S. E., Duren, R. M., Miller, C. E., Lai, C. T., Hsu, Y. K., Carranza, V., & Randerson, J. T. (2016). Mitigation of methane emissions in cities: How new measurements and partnerships can contribute to emissions reduction strategies. *Earth's Future, 4*, 408–425.

Howard, M., Hopkinson, P., & Miemczyk, J. (2019). The regenerative supply chain: A framework for developing circular economy indicators. *International Journal of Production Research, 57* (23), 7300–7318.

Huitema, D., Mostert, E., Egas, W., Moellenkamp, S., Pahl-Wostl, C., & Yalcin, R. (2009). Adaptive water governance: Assessing the institutional prescriptions of adaptive (co-) management from a governance perspective and defining a research agenda. *Ecology and Society, 14*(1), 26. Retrieved from http://www.ecologyandsociety.org/vol14/iss1/art26/.

Huntjens, P. (2011). *Water management and water governance in a changing climate. Experiences and insights on climate change adaptation in Europe, Africa, Asia and Australia.* Delft, the Netherlands: Eburon.

Huntjens, P. (2017). Mediation in the Israeli-Palestinian water conflict: A practitioner's view. In *Water diplomacy in action: Contingent approaches to managing complex water problems.* Anthem Press.

Huntjens, P. (2019, June). *Sociale innovatie voor een duurzame samenleving: Op weg naar een natuurlijk sociaal contract. Lectorale boek.* IMPACT Lectoraat Sociale Innovatie in het Groene Domein, Hogeschool Inholland.

Huntjens, P., & Nachbar, K. (2015). Climate change as a threat multiplier for human disaster and conflict. *Policy and Governance Recommendations for Advancing Climate Security, 4.*

Huntjens, P., & Zhang, T. (2016). *Climate justice: Equitable and inclusive governance of climate action.* The Hague Institute, Working Paper, 16.

Huntjens, P., Pahl-Wostl, C., & Grin, J. (2010). Climate change adaptation in European river basins. *Regional Environmental Change, 10*(4), 263–284.

Huntjens, P., Pahl-Wostl, C., Rihoux, B., Schlüter, M., Flachner, Z., Neto, S., Koskova, R., Nabide Kiti, I., & Dickens, C. (2011a). Adaptive water management and policy learning in a changing climate: A formal comparative analysis of eight water management regimes in Europe, Africa and Asia. *Environmental Policy and Governance, 21*(3), 145–163.

Huntjens, P. M. J. M., Termeer, C. J. A. M., Eshuis, J., & van Buuren, M. (2011b). *Collaborative action research for the governance of climate adaptation-foundations, conditions and pitfalls* (No. KfC/032/2010). Knowledge for Climate Programme Office.

Huntjens, P., Lebel, L., Pahl-Wostl, C., Camkin, J., Schulze, R., & Kranz, N. (2012). Institutional design propositions for the governance of adaptation to climate change in the water sector. *Global Environmental Change, 22*(1), 67–81.

Huntjens, P., Eshuis, J., Termeer, C. J. A. M., & van Buuren, M. W. (2014a). *Forms and foundations of action research.* London: Routledge.

Huntjens, P., Ottow, B., & Lasage, R. (2014b). Participation in climate adaptation in the Lower Vam Co River Basin in Vietnam. In *Action research for climate change adaptation* (pp. 71–91). Routledge.

Huntjens, P., De Man, R., Zhang, T., Van Rijswick, M., Misiedjan, D., Steenbergen, F., Evers, J., Al-Dawsari, N., Borgia, C., Tjen, A., Kwoei, A., Al-Kinda, A., & Al-Suneidar, M. (2014c). *The political economy of water management in Yemen: Conflict analysis and recommendations*. The Hague Institute for Global Justice. https://doi.org/10.13140/RG.2.2.16818.61122.

Huntjens, P., Valk, S., Zhang, T., & Warner, J. (2015, May). *Adaptive delta governance: Learning from dynamic deltas*. Policy Brief 15. The Hague Institute for Global Justice.

Huntjens, P., Yasuda, Y., Swain, A., de Man, R., Magsig, B. O., & Islam, S. (2016). *The multi-track water diplomacy framework: A legal and political economy analysis for advancing cooperation over shared waters* (1st ed.). The Netherlands: The Hague Institute for Global Justice.

Huntjens, P., Lebel, L., & Furze, B. (2017). The effectiveness of multi-stakeholder dialogues on water: Reflections on experiences in the Rhine, Mekong, and Ganga-Brahmaputhra-Meghna river basins. *International Journal of Water Governance, 5*(5:3), 39–60. https://doi.org/10.7564/15-IJWG98.

Huntjens, P., Zhang, T., & Nachbar, K. (2018). Climate change and implications for security and justice: The need for equitable, inclusive and adaptive governance of climate action. In B. Durch, R. Ponzio, & J. Larik (Eds.), *Just security in an undergoverned world*. Oxford University Press, 141 p.

Huntjens, P., Beers, P. J., Koot, J., & Wielinga, E. (2020). *The South Holland Food Family Programme: A transition approach towards a sustainable and self-sufficient food system*. Conference paper for the International Food and Agribusiness Management Association Conference 2020 (IFAMA2020), Rotterdam.

Hwang, A. (2000). Toward fostering systems learning in organizational contexts. *Systemic Practice and Action Research, 13*(3).

I&O Research. (2019). *Duurzaam denken is nog niet duurzaam doen: De CO_2 voetafdruk van Nederland*. I&O Research in opdracht van Binnenlands Bestuur.

Iansiti, M., & Levien, R. (2004). *The keystone advantage: What the new dynamics of business ecosystems mean for strategy, innovation, and sustainability*. Harvard Business Press.

Imperial, M. T. (1999). Institutional analysis and ecosystem-based management: The institutional analysis and development framework. *Environmental Management, 24*, 449–465.

Inglehart, R. (2017). Changing values in the Islamic world and the West. In *Values, political action, and change in the Middle East and the Arab Spring* (pp. 3–24). Oxford University Press.

Internal Displacement Monitoring Center. (2015). *Global estimates 2015: People displaced by disasters*. Geneva, Switzerland: Internal Displacement Monitoring Center.

IPBES. (2019, May 6). *Global assessment report on biodiversity and ecosystem services. Summary for policymakers*. Paris: Intergovernmental Science-Policy Platform on Biodiversity and Ecosystem Services (IPBES).

IPCC. (2013). Summary for policymakers. In *Climate change 2013: The physical science basis*. Cambridge: IPCC. Retrieved from http://www.ipcc.ch/pdf/assessment-report/ar5/wg1/WG1AR5_SPM_FINAL.pdf.

IPCC. (2014). *Contribution of working groups I, II and III to the fifth assessment report of the Intergovernmental Panel on Climate Change*. Climate change 2014: Synthesis report. Geneva: Intergovernmental Panel on Climate Change.

IPCC. (2018). Summary for policymakers. In V. Masson-Delmotte, P. Zhai, H.-O. Pörtner, D. Roberts, J. Skea, P. R. Shukla, A. Pirani, W. Moufouma-Okia, C. Péan, R. Pidcock, S. Connors, J. B. R. Matthews, Y. Chen, X. Zhou, M. I. Gomis, E. Lonnoy, M. T. Maycock, & T. Waterfield (Eds.), *Global warming of 1.5°C. An IPCC Special Report on the impacts of global warming of 1.5°C above pre-industrial levels and related global greenhouse gas emission pathways, in the context of strengthening the global response to the threat of climate change, sustainable development, and efforts to eradicate poverty*. Geneva, Switzerland: World Meteorological Organization, 32 p.

IPCC. (2019). *IPCC special report on climate change, desertification, land degradation, sustainable land management, food security, and greenhouse gas fluxes in terrestrial ecosystems* [A. Arneth, H. Barbosa, T. Benton, K. Calvin, E. Calvo, S. Connors, et al. (Eds.)]. IPCC.
Irwin, A. (1995). *Citizen science: A study of people, expertise and sustainable development*. London: Routledge.
Islam, S., & Madani, K. (Eds.). (2017). *Water diplomacy in action: Contingent approaches to managing complex water problems* (Vol. 1). Anthem Press.
Islam, M. T., & Nur, N. (2019). Climate change and violent conflict: An understanding and analysis of their causal relations. *Sub Journal of Sustainable Environment and Development, 1*.
Ison, R., Blackmore, C., & Iaquinto, B. L. (2013). Towards systemic and adaptive governance: Exploring the revealing and concealing aspects of contemporary social-learning metaphors. *Ecological Economics, 87*, 34–42.
Jackson, T. (2009). *Prosperity without growth: Economics for a finite planet*. Routledge.
Jahanbegloo, R. (1991). *Conversations with Isaiah Berlin*. McArthur & Co. Reprinted 2007, Halban Publishers. ISBN: 1-905559-03-8, ISBN: 978-1-905559-03-9.
Jänicke, M. (2012). "Green growth": From a growing eco-industry to economic sustainability. *Energy Policy, 48*, 13–21.
Janssen, C., & Erisman, J. W. (2016). Terug naar de bodem: Interview met Jan Willem Erisman in de Volkskrant: Groene landbouw: Biodiversiteit als beste basis voor het boerenbedrijf. *Volkskrant*.
Jennings, B. (2016). *Ecological governance: Toward a new social contract with the Earth*. Morgantown, WV: West Virginia University Press.
Johnson, N. (2018, February 9). *Doomsday debate: Wizards and prophets face off to save the planet*. Interview with Charles C. Mann. Retrieved from https://grist.org/article/wizards-and-prophets-face-off-to-save-the-planet/.
Jonker, J., Stegeman, H., & Faber, N. (2018). De circulaire economie: denkbeelden, ontwikkelingen en business modellen-2018 update.
Joshi, Y., & Rahman, Z. (2017). Investigating the determinants of consumers' sustainable purchase behaviour. *Sustainable Production and Consumption, 10*, 110–120.
JRC. (2019). *The future of cities: Opportunities, challenges and the way forward*. Joint Research Centre (JRC) of the European Commission. ISBN: 978-92-76-03847-4.
Kabisch, N., Korn, H., Stadler, J., & Bonn, A. (2017). *Nature-based solutions to climate change adaptation in urban areas: Linkages between science, policy and practice*. Springer Nature.
Kahane, A. (2004). *Solving tough problems: An open way of talking, listening, and creating new realities*. San Francisco: Berrett-Koehler.
Kallis, G., Kostakis, V., Lange, S., et al. (2018). Research on degrowth. *Annual Review of Environment and Resources, 43*, 291–316.
Kalshoven, F., & Zonderland, S. (2017). *Naar een nieuw sociaal contract: Voor een beter Nederland*. De Argumentenfabriek.
Kaplan, B., Kahn, L. H., & Monath, T. P. (2009). One health – One medicine: Linking human, animal and environmental health The brewing storm. *Veterinaria Italiana, 45*(1), 9–18.
Karré, P. M. (2018). Sociale innovatie bestuurskundig bekeken. In P. M. Karré, H. Dagevos, & G. Walraven (Eds.), *Sociale innovatie in de praktijk: Zoeken naar nieuwe antwoorden op maatschappelijke vraagstukken*. Koninklijke Van Gorcum.
Keizersrande. (2018). *Website Keizersrande*. Retrieved October 21, 2018, from http://www.keizersrande.nl.
Kemp, R., & Pearson, P. (2007). Final report MEI project about measuring eco-innovation. *UM Merit, Maastricht, 10*, 2.
Kemp, R., & Weehuizen, R. (2005). *Policy learning, what does it mean and how can we study it?* Publin Report No. D15, NIFU STEP, Oslo.
Kemp, R., Schot, J., & Hoogma, R. (1998). Regime shifts to sustainability through processes of niche formation: The approach of strategic niche management. *Technology Analysis & Strategic Management, 10*(2), 175–198.

Kemp, R., Rotmans, J., & Loorbach, D. (2007). Assessing the Dutch energy transition policy: How does it deal with dilemmas of managing transitions? *Journal of Environmental Policy Planning, 9*(34), 315–331.

Kern, K., & Alber, G. (2006). Governing climate change in cities: Modes of urban climate governance in multi-level systems. In *Proceedings OECD conference competitive cities and climate change* (p. 130). Paris: OECD.

Kern, F., & Smith, A. (2008). Restructuring energy systems for sustainability? Energy transition policy in the Netherlands. *Energy Policy, 36*(11), 4093–4103.

Kennedy, E., Gladek, E., & Roemers, G. (2018). *Using systems thinking to transform society.* https://www.metabolic.nl/publications/using-systems-thinking-totransform-society/

Kerschner, C. (2010). Economic de-growth vs. steady-state economy. *Journal of Cleaner Production, 18*(6), 544–551.

Keulartz, J., & Pekelharing, P. (2019, February 18). *Landbouw op de schop: Naar een duurzame voedselproductie.* Bureau De Helling.

Keyes, C. L. (2002). The mental health continuum: From languishing to flourishing in life. *Journal of Health and Social Behavior, 43*(2), 207–222.

Kirchherr, J., Reike, D., & Hekkert, M. (2017). Conceptualizing the circular economy: An analysis of 114 definitions. *Resources, Conservation and Recycling, 127*, 221–232.

Kirk, E., Orr, P., & Keyes, D. (2008). Environmental conflict resolution practice and performance: An evaluation framework. *Conflict Resolution Quarterly, 25*(3), 283–301.

Klein, N. (2015). *This changes everything: Capitalism vs. the climate.* Simon and Schuster.

Köhler, J., Geels, F. W., Kern, F., Markard, J., Onsongo, E., Wieczorek, A., Alkemade, F., Avelino, F., Bergek, A., Boons, F., & Fünfschilling, L. (2019). An agenda for sustainability transitions research: State of the art and future directions. *Environmental Innovation and Societal Transitions, 31*, 1–32.

Kondoh, K. (2015). The alternative food movement in Japan: Challenges, limits, and resilience of the teikei system. *Agriculture and Human Values, 32*(1), 143–153.

Konietzko, J., Bocken, N., & Hultink, E. J. (2020). Circular ecosystem innovation: An initial set of principles. *Journal of Cleaner Production, 253*, 119942.

Kooiman, J. (Ed.). (1993). *Modern governance: New government-society interactions.* London: Sage Publications.

Korhonen, J., Nuur, C., Feldmann, A., & Birkie, S. E. (2018). Circular economy as an essentially contested concept. *Journal of Cleaner Production, 175*, 544–552.

Krauß, W., Bremer, S., Wardekker, J. A., Marschütz, B., Baztan, J., & da Cunha, C. (2018). *Initial mapping of narratives of change.*

Kremer, M., & Maskin, E. (2006). *Globalization and inequality.* Harvard University Press.

Krenek, A., & Schratzenstaller, M. (2016). *Sustainability-oriented EU taxes: The example of a European carbon-based flight ticket tax.* WIFO.

Kübler, D. (1999). Ideas as catalytic elements for policy change: Advocacy coalitions and drug policy in Switzerland. In D. Braun & A. Busch (Eds.), *Public policy and political ideas* (pp. 116–135). Cheltenham: Edward Elgar Publishing Limited.

Laaksonen, S., Pusenius, J., Kumpula, J., Venäläinen, A., Kortet, R., Oksanen, A., & Hoberg, E. (2010). Climate change promotes the emergence of serious disease outbreaks of filarioid nematodes. *EcoHealth, 7*(1), 7–13.

LaCanne, C. E., & Lundgren, J. G. (2018). Regenerative agriculture: Merging farming and natural resource conservation profitably. *PeerJ, 6*, e4428.

Lansing, J. S. (2003). Complex adaptive systems. *Annual Review of Anthropology, 32*(1), 183–204.

Larrubia Vargas, R. (2017). The common agricultural policy and its reforms: Reflections about the 2014–2020 reform. *Cuadernos Geográficos, 56*(1), 124–147.

Lasswell, H. D. (1950). *Politics: Who gets what, when, how.* New York: P. Smith.

Latour, B. (2012). *We have never been modern.* Harvard University Press.

Latour, B. (2018, November). *Waar kunnen we landen? Politieke oriëntatie in het Nieuwe Klimaatregime* (1st ed.). Octavo Publicaties, 120 p.

Lave, J., & Wenger, E. (1991). *Situated learning: Legitimate peripheral participation*. Cambridge, UK: Cambridge University Press. First published in 1990 as Institute for Research on Learning report 90-0013.

Lawrence, R. L., Daniels, S. E., & Stankey, G. H. (1997). *Procedural justice and public involvement in natural resource decision making*.

Lawrence, T. B., Suddaby, R., & Leca, B. (Eds.). (2009). *Institutional work: Actors and agency in institutional studies of organizations*. Cambridge University Press.

Lax, D. A., & Sebenius, J. K. (2006). Making lasting deals. In *3D negotiation: Powerful tools to change the game in your most important deals* (pp. 149–161). Boston, MA: Harvard Business School Press.

Lay, B., Neyret, L., Short, D., Baumgartner, M. U., & Oposa, A. A., Jr. (2015). Timely and necessary: Ecocide law as urgent and emerging. *The Journal Jurisprudence, 28*, 431.

Leach, W. D., & Pelkey, N. W. (2001). Making watershed partnerships work: A review of the empirical literature. *Journal of Water Resources Planning and Management, 127*, 378–385.

Leach, M., Bloom, G., Ely, A., Nightingale, P., Scoones, I., Shah, E., & Smith, A. (2007). *Understanding governance: Pathways to sustainability*. STEPS Centre.

Lebel, L., Anderies, J. M., Campbell, B., Folke, C., Hatfield-Dodds, S., Hughes, T. P., & Wilson, J. (2006). Governance and the capacity to manage resilience in regional social-ecological systems. *Ecology and Society, 11*(1).

Lee, K. N. (1999). Appraising adaptive management. *Conservation Ecology, 3*(2), 3.

Lee, S., Park, G., Yoon, B., & Park, J. (2010). Open innovation in SMEs—An intermediated network model. *Research Policy, 39*(2), 290–300.

Leggewie, C., & Messner, D. (2012). The low-carbon transformation—A social science perspective. *Journal of Renewable and Sustainable Energy, 4*, 041404.

Leicester, G. (2007). Policy learning: Can government discover the treasure within? *European Journal of Education, 42*(2), 173–184. Published Online: 23 May 2007.

Levin, S. A. (1998). Ecosystems and the biosphere as complex adaptive systems. *Ecosystems, 1*(5), 431–436.

Levy, P. (2003). A methodological framework for practice-based research in networked learning. *Instructional Science, 31*, 87–109.

Lieder, M., & Rashid, A. (2016). Towards circular economy implementation: A comprehensive review in context of manufacturing industry. *Journal of Cleaner Production, 115*, 36–51.

Lintsen, H., Veraart, F., Smits, P. J., & Grin, J. (2008, March 14). *De kwetsbare welvaart van Nederland, 1850–2050: Naar een circulaire economie*. Prometheus, 560 p.

Liu, Y., Sheng, H., Mundorf, N., Redding, C., & Ye, Y. (2017). Integrating norm activation model and theory of planned behavior to understand sustainable transport behavior: Evidence from China. *International Journal of Environmental Research and Public Health, 14*(12), 1593.

LNV. (2018, September 8). *Visie Landbouw, Natuur en Voedsel: Waardevol en Verbonden*. Beleidsnota Ministerie Landbouw, Natuur en Voedselkwaliteit.

Loeber, A. (2003). *Inbreken in het gangbare. Transitiemanagement in de praktijk: de NIDO-benadering*. Leeuwarden: Nationaal Initiatief Duurzame Ontwikkeling.

Logatscheva, K. (2016). *Monitor Duurzaam Voedsel*. Wageningen Universiteit.

Logatscheva, K. (2019). *Monitor Duurzaam Voedsel*. Wageningen Universiteit.

London, T. (2009). Making better investments at the base of the pyramid. *Harvard Business Review, 87*(5), 106–113.

Loorbach, D. (2010). Transition management for sustainable development: a prescriptive, complexity-based governance framework. *Governance, 23*(1), 161–183.

Loorbach, D., & Rotmans, J. (2006). *Managing transitions for sustainable development*. International Center for Integrative Studies, Maastricht University.

Loorbach, D., Frantzeskaki, N., & Avelino, F. (2017). Sustainability transitions research: Transforming science and practice for societal change. *Annual Review of Environment and Resources, 42*, 599–626.

Luederitz, C., Schapke, N., Wiek, A., Lang, D. J., Bergmann, M., Bos, J. J., Burch, S., Davies, A., Evans, J., Konig, A., Farrelly, M. A., Forrest, N., Frantzeskaki, N., Gibson, R. B., Kay, B., Loorbach, D., McCormick, K., Parodi, O., Rauschmayer, F., Schneidewind, U., Stauffacher, M., Stelzer, F., Trencher, G., Venjakob, J., Vergragt, P. J., von Wehrden, H., & Westley, F. R. (2017). Learning through evaluation – A tentative evaluative scheme for sustainability transition experiments. *Journal of Cleaner Production, 169*, 61–76.

Maani, K., & Cavana, R. Y. (2007). *Systems thinking, system dynamics: Managing change and complexity*. Prentice Hall.

Maas, T., van den Broek, J., & Deuten, J. (2017). *Living labs in Nederland: Van open testfaciliteit tot levend lab*. Den Haag: Rathenau Instituut.

Main, A. R. (1999). How much biodiversity is enough? *Agroforestry Systems, 45*(1–3), 23–41.

Mann, C. C. (2018). *The Wizard and the Prophet: Two remarkable scientists and their dueling visions to shape tomorrow's world*. Knopf.

Manomaivibool, P., Chart-asa, C., & Unroj, P. (2016). Measuring the impacts of a save food campaign to reduce food waste on campus in Thailand. *Applied Environmental Research, 38*, 13–22. https://doi.org/10.35762/AER.2016.38.2.2.

Martella, J., Rendon, T., & Schilder, D. (2019, June 3). *System leaders and system thinkers*. Presentation by NAEYC Professional Learning Institute.

Martin, A., & Sutherland, A. (2003). Whose research, whose agenda? In B. Pound, S. Snapp, C. McDougall, & A. Braun (Eds.), *Managing natural resources for sustainable livelihoods – Uniting science and participation*. Earthscan/IDRC.

Martínez-Alier, J., Pascual, U., Vivien, F. D., & Zaccai, E. (2010). Sustainable de-growth: Mapping the context, criticisms and future prospects of an emergent paradigm. *Ecological Economics, 69*(9), 1741–1747.

Mason, P. (2016). *Postcapitalism: A guide to our future*. Macmillan.

Mathews, F. (2011). Towards a deeper philosophy of biomimicry. *Organization & Environment, 24*(4), 364–387.

Maurer, M., Bell, E., Woods, E., & Allen, E. (2006, December). Structured discovery in cane travel: Constructivism in action. *Phi Delta Kappan, 88*(4), 304–307.

Mazziotta, M., & Pareto, A. (2013). A non-compensatory composite index for measuring well-being over time. *Cogito. Multidisciplinary Research Journal, 5*(4), 93–104.

Mazzucato, M. (2018). *The value of everything: Making and taking in the global economy*. Hachette UK.

McCauley, D. J. (2006). Selling out on nature. *Nature, 443*(7), 27–28.

McDonald, M. (March 2013). Discourses of climate security. *Political Geography, 33*, 42–51.

McGinnis, M. (2000). *Polycentric governance and development*. Ann Arbor, MI: University of Michigan Press.

McKibben, B. (2019, April 16). *Falter: Has the human game begun to play itself out?* Henry Holt and Co, ISBN-10: 1250178266.

Meadowcroft, J. (2007). Who is in charge here? Governance for sustainable development in a complex world. *Journal of Environmental Policy Planning, 9*(3–4), 299–314.

Meadowcroft, J. (2009). What about the politics? Sustainable development, transition management, and long term energy transitions. *Policy Sciences, 42*(4), 323–340.

Meadows, D. H. (2008). *Thinking in systems: A primer*. Hartford, VT: Chelsea Green Publishing.

Mendelberg, T. (2002). The deliberative citizen: Theory and evidence. *Political Decision Making, Deliberation and Participation, 6*(1), 151–193.

Mercier, H., & Landemore, H. (2012). Reasoning is for arguing: Understanding the successes and failures of deliberation. *Political Psychology, 33*(2), 243–258.

Millennium Ecosystem Assessment. (2005). *Ecosystems and human well-being: Biodiversity synthesis*. Washington, DC: World Resources Institute.

Mitchell, R. (2013). Is physical activity in natural environments better for mental health than physical activity in other environments? *Social Science & Medicine, 91*, 130–134.

Mnookin, R., Pepper, S., & Tulumello, A. (2000). The testion between creating and distributing value. In *Beyond winning: Negotiating to create value in deals and disputes* (pp. 11–43). Cambridge, MA: Harvard University Press.

Mohajerani, A., Bakaric, J., & Jeffrey-Bailey, T. (2017). The urban heat island effect, its causes, and mitigation, with reference to the thermal properties of asphalt concrete. *Journal of Environmental Management, 197*, 522–538.

Mommers, J. (2019). Hoe gaan we dit uitleggen: onze toekomst op een steeds warmere aarde. De Correspondent.

Moon, H. C., & Parc, J. (2019). Shifting corporate social responsibility to corporate social opportunity through creating shared value. *Strategic Change, 28*(2), 115–122.

Moon, H. C., Parc, J., Yim, S. H., & Park, N. (2011). An extension of Porter and Kramer's Creating Shared Value (CSV): Reorienting strategies and seeking international cooperation. *Journal of International and Area Studies, 18*(2), 49–64.

Moreno, M., & Charnley, F. (2016). Can re-distributed manufacturing and digital intelligence enable a regenerative economy? An integrative literature review. In *International conference on sustainable design and manufacturing* (pp. 563–575). Cham: Springer.

Morgan, K. (2009). *Feeding the city: The challenge of urban food planning*.

Morris, C., Kirwan, J., & Lally, R. (2014). Less meat initiatives: An initial exploration of a diet-focused social innovation in transitions to a more sustainable regime of meat provisioning. *International Journal of Sociology of Agriculture and Food, 21*, 189–208.

Moulaert, F., MacCallum, D., & Hillier, J. (2013). Social innovation: Intuition, precept, concept, theory and practice. In F. Moulaert, D. MacCallum, A. Mehmood, & A. Hamdouch (Eds.), *The international handbook of social innovation: Collective action, social learning and transdisciplinary research* (pp. 13–24). Cheltenham: Edward Elgar Publishing.

Muro, M., & Jeffrey, P. (2008). A critical review of the theory and application of social learning in participatory natural resources management. *Journal of Environmental Planning and Management, 51*(3), 325–344.

Murray, A., Skene, K., & Haynes, K. (2017). The circular economy: An interdisciplinary exploration of the concept and application in a global context. *Journal of Business Ethics, 140*(3), 369–380.

Myers, J. (2019, August 28). *This man is turning cities into giant sponges to save lives*. World Economic Forum. Retrieved from https://www.weforum.org/agenda/2019/08/sponge-cities-china-flood-protection-nature-wwf/.

Nash, R. F. (1989). *The rights of nature: A history of environmental ethics*. Madison, WI: University of Wisconsin Press. ISBN: 978-0299118440.

Nefs, M. (2017). 80% van Zuid-Hollandse maaltijden van lokale bodem – kan dat? *Verse Stad*. Retrieved March 6, 2019, from https://versestad.nl/2017/02/80-van-zuid-hollandse-maaltijden-van-lokale-bodem-kan-dat/.

Negowetti, N. E. (2016). Exposing the invisible costs of commercial agriculture: Shaping policies with true costs accounting to create a sustainable food future. *Valparaiso University Law Review, 51*, 447.

NewForesight and Commonland. (2017). *New horizons for the transitioning of our food system: Connecting ecosystems, value chains and consumers*. Discussion paper by NewForesight and Commonland with contributions from The Boston Consulting Group.

Newman, L. L., & Dale, A. (2005). Network structure, diversity, and proactive resilience building: A response to Tompkins and Adger. *Ecology and Society, 10*(1), r2. [Online]. Retrieved from http://www.ecologyandsociety.org/vol10/iss1/resp2/.

Newman, P., Beatley, T., & Boyer, H. (2017). Foster inclusive and healthy cities. In *Resilient cities* (pp. 89–106). Washington, DC: Island Press.

Oates, W. E. (1998). Environmental policy in the European community: Harmonization or national standards? *Empirica, 25*, 1–13.

O'Brien, R. (2001). An overview of the methodological approach of action research. In R. Richardson (Ed.), *Theory and practice of action research*. João Pessoa, Brazil: Universidade Federal da Paraíba.

O'Brien, K. (2012). Global environmental change II: From adaptation to deliberate transformation. *Progress in Human Geography, 36*, 667–676.

O'Brien, K., Hayward, B., & Berkes, F. (2009). Rethinking social contracts: Building resilience in a changing climate. *Ecology and Society, 14*, 12.

OECD. (2011). *Towards green growth*. Paris, France: OECD Publishing.

Olsen, J. P. (2009). Change and continuity: An institutional approach to institutions of democratic government. *European Political Science Review, 1*, 3–12.

Olsson, P., & Galaz, V. (2012). Social-ecological innovation and transformation. In *Social innovation* (pp. 223–247). London: Palgrave Macmillan.

Olsson, P., Folke, C., & Berkes, F. (2004). Adaptive comanagement for building resilience in social–ecological systems. *Environmental Management, 34*(1), 75–90.

O'Neill, D. W. (2012). Measuring progress in the degrowth transition to a steady state economy. *Ecological Economics, 84*, 221–231.

Onwezen, M. C., Antonides, G., & Bartels, J. (2013). The Norm Activation Model: An exploration of the functions of anticipated pride and guilt in pro-environmental behaviour. *Journal of Economic Psychology, 39*, 141–153.

Osborne, S. P. (2009, January). *An introduction to game theory* (1st ed.). Uitgever: Atheneum Uitgeverij, Engels, Paperback, 560 p.

Osborne, S. P. (2010). Introduction – The (new) public governance: A suitable case for treatment? In S. P. Osborne (Ed.), *The new public governance? Emerging perspectives on the theory and practice of public governance* (pp. 1–16). Abingdon: Routledge.

Ostrom, E. (1990). *Governing the commons: The evolution of institutions for collective action*. Cambridge University Press.

Ostrom, E. (1996). Crossing the great divide: Coproduction, synergy, and development. *World Development, 24*(6), 1073–1087.

Ostrom, E. (1999). Institutional rational choice: An assessment of the institutional analysis and development framework. In P. A. Sabatier (Ed.), *Theories of the policy process*. Oxford: Westview Press.

Ostrom, E. (2005). *Understanding institutional diversity*. Princeton: Princeton University Press.

Ostrom, E. (2007). A diagnostic approach for going beyond Panaceas. *Proceedings of the National Academy of Sciences, 104*, 15181–15187.

Ostrom, E. (2009). *Understanding institutional diversity*. Princeton, NJ: Princeton University Press.

Otero, G., Pechlaner, G., & Gürcan, E. C. (2013). The political economy of "food security" and trade: Uneven and combined dependency. *Rural Sociology, 78*(3), 263–289.

Pahl-Wostl, C. (1995). *The dynamic nature of ecosystems: Chaos and order entwined*. Chichester: Wiley & Sons.

Pahl-Wostl, C. (2007). Transition towards adaptive management of water facing climate and global change. *Water Resources Management, 21*, 49–62.

Pahl-Wostl, C. (2009). A conceptual framework for analysing adaptive capacity and multi-level learning processes in resource governance regimes. *Global Environmental Change, 19*(3), 354–365.

Pahl-Wostl, C., Craps, M., Dewulf, A., Mostert, E., Tabara, D., & Taillieu, T. (2007). Social learning and water resources management. *Ecology and Society, 12*, 5.

Pahl-Wostl, C., Bhaduri, A., & Gupta, J. (Eds.). (2016). *Handbook on water security*. Edward Elgar Publishing.

Parker, J., Mars, L., Ransome, P., & Stanworth, H. (2003). *Social theory: A basic tool kit*. Hampshire: Palgrave Macmillan.

Patel, R., & Moore, J. W. (2017). *A history of the world in seven cheap things: A guide to capitalism, nature, and the future of the planet*. Oakland, CA: Univ of California Press.

Peek, G. J. (2015). *Stedelijke gebiedsontwikkeling in transitie* (1st ed.). Hogeschool Rotterdam Uitgeverij. ISBN: 90-5179-921-7.

Pelling, M. (2010). *Adaptation to climate change: From resilience to transformation*. Routledge.

Peters, J., & Weggeman, M. C. D. P. (2009). *Het Rijnland boekje: principes en inzichten van het Rijnland-model*. Amsterdam: Business Contact. ISBN: 978-90-470-0209-3.

Petridis, P., Muraca, B., & Kallis, G. (2015). Degrowth: Between a scientific concept and a slogan for a social movement. In *Handbook of ecological economics*. Edward Elgar Publishing.

Pharo, P., et al. (2019). *Growing better: Ten critical transitions to transform food and land use*. FOLU.

Pigford, A. A. E., Hickey, G. M., & Klerkx, L. (2018). Beyond agricultural innovation systems? Exploring an agricultural innovation ecosystems approach for niche design and development in sustainability transitions. *Agricultural Systems, 164*, 116–121.

Piketty, T. (2013). *Le capital au XXIe siècle*. Le Seuil.

Pikramenou, N. (2020). *Rights of nature: Time to shift the paradigm in the EU?* Accessed on 11 November 2020 at Earth Law Center. https://www.earthlawcenter.org/nikolettas-ron-article.

Pirson, M. (2012). Social entrepreneurs as the paragons of shared value creation? A critical perspective. *Social Enterprise Journal, 8*(1).

Polanyi, K. (1944). *The great transformation*. New York: Rinehart.

Polonsky, M., Bhaskaran, S., Cary, J., & Fernandez, S. (2006). Environmentally sustainable food production and marketing: Opportunity or hype? *British Food Journal, 108*(8), 677–690.

Polsby, N. W. (1984). *Political innovation in America: The politics of policy initiation*. New Haven: Yale University Press.

Porter, M. E., & Kramer, M. R. (2002). The competitive advantage of corporate. *Harvard Business Review, 80*(12), 56–69.

Porter, M. E., & Kramer, M. R. (2019). Creating shared value. In *Managing sustainable business* (pp. 323–346). Dordrecht: Springer.

Porter, M., Hills, G., Pfitzer, M., Patscheke, S., & Hawkins, E. (2012). *Measuring shared value: How to unlock value by linking social and business results*. FSG. Retrieved from http://www.fsg.org/tabid/191/ArticleId/740/Default.aspx?srpush=true.

Pötz, H., & Bleuzé, P. (2016). *Groenblauwe netwerken. Handleiding voor veerkrachtige steden*. Uitgever: atelier GROENBLAUW. ISBN: 978-90-9029-822-1.

Power, E. M. (1999). Combining social justice and sustainability for food security. For hunger-proof cities sustainable urban food systems (pp. 30–37). Ottawa: International Development Research Centre.

Prendeville, S., Cherim, E., & Bocken, N. (2018). Circular cities: Mapping six cities in transition. *Environmental Innovation and Societal Transitions, 26*, 171–194.

Prieto-Sandoval, V., Jaca, C., & Ormazabal, M. (2018). Towards a consensus on the circular economy. *Journal of Cleaner Production, 179*, 605–615.

Puente-Rodríguez, D., Bos, A. B., & Koerkamp, P. W. G. (2019). Rethinking livestock production systems on the Galápagos Islands: Organizing knowledge-practice interfaces through reflexive interactive design. *Environmental Science & Policy, 101*, 166–174.

Puerari, E., De Koning, J. I., Von Wirth, T., Karré, P. M., Mulder, I. J., & Loorbach, D. A. (2018). Co-creation dynamics in urban living labs. *Sustainability, 10*(6), 1893.

Quist, J., & Vergragt, P. (2006). Past and future of backcasting: The shift to stakeholder participation and a proposal for a methodological framework. *Futures, 38*(9), 1027–1045. ISSN: 0016-3287. https://doi.org/10.1016/j.futures.2006.02.010.

Raiffa, H. (1982). Analytical models and empirical results. In *The art and science of negotiation* (pp. 44–65). Cambridge, MA: Harvard University Press.

Raven, R., Van den Bosch, S., & Weterings, R. (2010). Transitions and strategic niche management: Towards a competence kit for practitioners. *International Journal of Technology Management, 51*(1), 57–74.

Raworth, K. (2017). *Doughnut economics, seven ways to think like a 21st century economist*. New Orleans, LA: Cornerstone.

Reason, P., & Bradbury, H. (Eds.). (2005). *Handbook of action research: Concise paperback edition*. London: Sage.

Reed, M. S. (2008). Stakeholder participation for environmental management: A literature review. *Biological Conservation, 141*, 2417–2431.

Rees, E. E., Ng, V., Gachon, P., Mawudeku, A., McKenney, D., Pedlar, J., et al. (2019). Climate change and infectious diseases: The solutions: Risk assessment strategies for early detection and prediction of infectious disease outbreaks associated with climate change. *Canada Communicable Disease Report, 45*(5), 119.

Regeer, B. J., Hoes, A. C., van Amstel-van Saane, M., Caron-Flinterman, F., & Bunders, J. (2009). Six guiding principles for evaluating mode-2 strategies for sustainable development. *American Journal of Evaluation, 30*(4), 515–537.

Reicher, S., & Hopkins, N. (2001). *Self and nation: Categorization, contestation and mobilization*. London, UK: SAGE.

Rhodes, C. J. (2017). The imperative for regenerative agriculture. *Science Progress, 100*(1), 80–129.

Rijksoverheid2. (n.d.). Landbouw, voedsel, natuur, visserij en dierenwelzijn. Retrieved from March 7, 2019, from https://www.rijksoverheid.nl/regering/regeerakkoord-vertrouwen-in-de-toekomst/3.-neder-land-wordt-duurzaam/3.4-landbouw-voedsel-natuur-visserij-en-dierenwelzijn.

Rip, A., Voss, J. P., & Bauknecht, D. (2006). A co-evolutionary approach to reflexive governance – And its ironies. *Reflexive Governance for Sustainable Development*, 82–100.

Rissman, A. R., & Gillon, S. (2017). Where are ecology and biodiversity in social–ecological systems research? A review of research methods and applied recommendations. *Conservation Letters, 10*(1), 86–93.

Rittel, H., & Webber, M. (1973). Dilemmas in a general theory of planning. *Policy Sciences, 4*, 155–169.

RNE. (2016, November). *Roadmap next economy*. Metropoolregio Rotterdam Den Haag. Retrieved from www.mrdh.nl/rne.

Rockström, J., & Karlberg, L. (2010). The Quadruple Squeeze: Defining the safe operating space for freshwater use to achieve a triply green revolution in the Anthropocene. *Ambio, 39*(3), 257–265.

Rockstrom, J., Steffen, W., Noone, K., Persson, A., Chapin, F. S., III, Lambin, E., Lenton, T. M., Scheffer, M., Folke, C., Schellnhuber, H., Nykvist, B., De Wit, C. A., Hughes, T., van der Leeuw, S., Rodhe, H., Sorlin, S., Snyder, P. K., Costanza, R., Svedin, U., Falkenmark, M., Karlberg, L., Corell, R. W., Fabry, V. J., Hansen, J., Walker, B., Liverman, D., Richardson, K., Crutzen, P., & Foley, J. (2009). Planetaryboundaries: Exploring the safe operating space for humanity. *Ecology and Society, 14*(2), 32. [Online]. Retrieved from http://www.ecologyandsociety.org/vol14/iss2/art32/.

Rockström, J., Edenhofer, O., Gaertner, J., et al. (2020). Planet-proofing the global food system. *Nature Food, 1*, 3–5. https://doi.org/10.1038/s43016-019-0010-4.

Rodríguez-Carvajal, R., Moreno-Jiménez, B., de Rivas-Hermosilla, S., Álvarez-Bejarano, A., & Vergel, A. I. S. (2010). Positive psychology at work: Mutual gains for individuals and organizations. *Revista de Psicología del Trabajo y de las Organizaciones, 26*(3), 235–253.

Rojas, J. (2014). *The impact of capitalism and materialism on generosity: A cross-national examination*. PhD-thesis, University of Iowa. https://doi.org/10.17077/etd.z5vnrer4.

Romme, A. G. L., & Van Witteloostuijn, A. (1999). Circular organizing and triple loop learning. *Journal of Organizational Change Management, 12*(5), 439–454.

Rosen, A. (2015, February). The wrong solution at the right time: The failure of the Kyoto protocol on climate change. *Politics & Policy, 43*(1), 30–58.

Rotjan, R. D., Chabot, J. R., & Lewis, S. M. (2010). Social context of shell acquisition in Coenobita clypeatus hermit crabs. *Behavioral Ecology, 21*(3), 639–646.

Rotmans, J. (2005). *Societal innovation: Between dream and reality lies complexity*. Inaugural Speech, Booklet, Erasmus University Rotterdam.

Rotmans, J. (2010, June). *Transitieagenda voor Nederland: Investeren in duurzame innovatie*. Uitgave van Kennisnetwerk Systeeminnovaties en transities (KSI) p/a DRIFT, ISBN: 978-90-9025499-9.

Rotmans, J. (2019). *Review op Expertpaper over het bewerkstelligen van een transitie naar kringlooplandbouw*. Termeer C.J.A.M.

Rotmans, J., Kemp, R., & van Asselt, M. (2001). More evolution than revolution: Transition management in public policy. *Foresight, 3*(1), 15–31.

Rowe, P. G. (1987). *Design thinking*. MIT Press.

Rübbelke, D., & Vögele, S. (2013). Short-term distributional consequences of climate change impacts on the power sector: Who gains and who loses? *Climatic Change, 116*(2), 191–206.

Rühs, N., & Jones, A. (2016). The implementation of Earth jurisprudence through substantive constitutional rights of nature. *Sustainability, 8*(2), 174.

Runhaar, H. (2017). Governing the transformation towards 'nature-inclusive' agriculture: Insights from the Netherlands. *International Journal of Agricultural Sustainability, 15*(4), 340–349.

Runhaar, H. A. C., Melman, T. C. P., Boonstra, F. G., Erisman, J. W., Horlings, L. G., de Snoo, G. R., Termeer, C. J. A. M., Wassen, M. J., Westerink, J., & Arts, B. J. M. (2017). Promoting nature conservation by Dutch farmers: A governance perspective. *International Journal of Agricultural Sustainability, 15*(3), 264–281. https://doi.org/10.1080/14735903.2016.1232015.

Runhaar, H., Fünfschilling, L., van den Pol-Van Dasselaar, A., Moors, E. H. M., Temmink, R., & Hekkert, M. (2020). Endogenous regime change: Lessons from transition pathways in Dutch dairy farming. *Environmental Innovation and Societal Transitions, 36*, 137–150. https://doi.org/10.1016/j.eist.2020.06.001.

Russell, R., Guerry, A. D., Balvanera, P., Gould, R. K., Basurto, X., Chan, K. M., Klain, S., Levine, J., & Tam, J. (2013). Humans and nature: how knowing and experiencing nature affect well-being. *Annual Review of Environment and Resources, 38*, 473–502.

Russett, B. M. (1964). Inequality and instability: The relation of land tenure to politics. *World Politics, 16*(3), 442–454.

Ryan, L., & Wallace, J. (2019). Mutual gains success and failure: Two case studies of annual hours in Ireland. *The Irish Journal of Management, 1*(ahead-of-print).

Ryszawska, B. (2018). Role of banks in sustainable and digital transition. *Financial Sciences. Nauki o Finansach, 23*(1), 65–74.

Sabatier, P. A. (1988). An advocacy coalition framework of policy change and the role of policy-oriented learning therein. *Policy Sciences, 21*(Fall), 129–168.

Sabatier, P. A., & Jenkins-Smith, H. C. (Eds.). (1993). *Policy change and learning: An advocacy coalition approach*. Boulder: Westview Press.

Sabatier, P. A., & Jenkins-Smith, H. C. (1999). The advocacy coalition framework: An assessment. In P. A. Sabatier (Ed.), *Theories of the policy process*. Boulder: Westview Press.

Sandel, M. (2012, May). *Niet alles is te koop: de morele grenzen van marktwerking* (1st ed.). Uitgeverij Ten Have, 288 p.

Sandel, M., Macintyre, A., Barber, B., & Taylor, C. (1985). Liberalism and the limits of justice. *Philosophy and Public Affairs, 14*(3), 308–322.

Sanders, M. E., Westerink, J., Migchels, G., Korevaar, H., Geerts, R. H. E. M., Bloem, J., et al. (2015). *Op weg naar een natuurinclusieve duurzame landbouw*. Wageningen-UR: Alterra.

Sanderson, I. (2002). Evaluation, policy learning and evidence-based policy making. *Public Administration, 80*(1), 1–22.

Sandifer, P. A., Sutton-Grier, A. E., & Ward, B. P. (2015). Exploring connections among nature, biodiversity, ecosystem services, and human health and well-being: Opportunities to enhance health and biodiversity conservation. *Ecosystem Services, 12*, 1–15.

Santha, S. D. (2020). *Climate change and adaptive innovation: A model for social work practice*. Routledge.

Santos, F. M., & Eisenhardt, K. M. (2005). Organizational boundaries and theories of organization. *Organization Science, 16*(5), 491–508.

SAPEA, Science Advice for Policy by European Academies. (2020). *A sustainable food system for the European Union: A systematic review of the European policy ecosystem*. Berlin: SAPEA. https://doi.org/10.26356/sustainablefoodreview.
Sassen, S. (2014). *Expulsions: Brutality and complexity in the global economy*. Harvard University Press.
Scharmer, C. O., & Kaufer, K. (2013). *Leading from the emerging future: From ego-system to eco-system economies*. San Francisco, CA: Berrett-Koehler Publishers.
Schellnhuber, H. J., Messner, D., Leggewie, C., Leinfelder, R., Nakicenovic, N., Rahmstorf, S., et al. (2011). *World in transition: A social contract for sustainability*. Flagship Report.
Scherb, A., Palmer, A., Frattaroli, S., & Pollack, K. (2012). Exploring food system policy: A survey of food policy councils in the United States. *Journal of Agriculture, Food Systems, and Community Development, 2*(4), 3–14.
Scheve, K., & Stasavage, D. (2016). *Taxing the rich: A history of fiscal fairness in the United States and Europe*. Princeton University Press.
Schipanski, M. E., MacDonald, G. K., Rosenzweig, S., Chappell, M. J., Bennett, E. M., Kerr, R. B., et al. (2016). Realizing resilient food systems. *BioScience, 66*(7), 600–610.
Schmitt, G., Tapias, E., & Wisniewska, H. M. (2019). *City in your hands*. Swiss Federal Institute of Technology in Zurich (ETHZ), Department of Architecture, Chair of Information Architecture.
Schneider, F., Kallis, G., & Martinez-Alier, J. (2010). Crisis or opportunity? Economic degrowth for social equity and ecological sustainability. Introduction to this special issue. *Journal of Cleaner Production, 18*(6), 511–518.
Schot, J., & Geels, F. W. (2008). Strategic niche management and sustainable innovation journeys: Theory, findings, research agenda, and policy. *Technology Analysis & Strategic Management, 20*(5), 537–554.
Schouten, C., & Spijker, G. J. (2017). *Rijnland werkt: In gesprek met ondernemers* (Christelijk-sociaal 2030, Deel 7). Wetenschappelijk Instituut ChristenUnie. ISBN: 978-90-5881-938-3.
Schwartz, S. H. (1977). Normative influences on altruism. In L. Berkowitz (Ed.), *Advances in experimental social psychology* (Vol. 10, pp. 221–279). New York: Academic Press.
Schwartz, S. H., & Bilsky, W. (1987). Toward a universal psychological structure of human values. *Journal of Personality and Social Psychology, 53*(3), 550.
Schwartz, S. H., & Howard, J. A. (1981). A normative decision-making model of altruism. In J. P. Rushton & R. M. Sorrentino (Eds.), *Altruism and helping behavior: Social, personality, and developmental perspectives* (pp. 189–211). Hillsdale: Lawrence Erlbaum.
Seo, M. G., & Creed, W. D. (2002). Institutional contradictions, praxis, and institutional change: A dialectical perspective. *Academy of Management Review, 27*(2), 222–247.
Shove, E., & Walker, G. (2007). Caution! Transition ahead: Policies, practice, and sustainable transition management. *Environment and Planning, 39*, 763–770.
Sieveking, A. (2019). Food policy councils as loci for practising food democracy? Insights from the case of Oldenburg, Germany. *Politics and Governance, 7*(4), 48–58.
Singh, V. P. (2017). Challenges in meeting water security and resilience. *Water International, 42*(4), 349–359.
Skandrani, Z. (2016). Considering the socio-ecological co-construction of nature conceptions as a basis for urban environmental governance. *Journal of Geography and Natural Disasters, 6*(1), 1–3.
SKIN. (2020). *Short supply chains*. Knowledge & Innovation Network (SKIN). Retrieved May 28, 2020, from http://www.shortfoodchain.eu/.
Smink, M. (2015). *Incumbents and institutions in sustainability transitions*. Utrecht University.
Snyder, C. R., Lopez, S. J., & Pedrotti, J. T. (2011). *Positive psychology: The scientific and practical explorations of human strengths*. Los Angeles: Sage.
Sociaal en Cultureel Planbureau. (2016, January). *Kiezen bij de kassa: Een verkenning van maatschappelijk bewust consumeren in Nederland*. SCP.

Sol, J., van der Wal, M. M., Beers, P.-J., & Wals, A. E. J. (2018). Reframing the future: The role of reflexivity in governance networks in sustainability transitions. *Environmental Education Research, 24*(9), 1383–1405.

Sovacool, B. K. (2016). *The history and politics of energy transitions: Comparing contested views and finding common ground (No. 2016/81)*. WIDER Working Paper.

Spangenberg, J. H. (2005). Economic sustainability of the economy: Concepts and indicators. *International Journal of Sustainable Development, 8*(1–2), 47–64.

Sparkman, G., & Walton, G. M. (2017). Dynamic norms promote sustainable behavior, even if it is counternormative. *Psychological Science, 28*(11), 1663–1674. https://doi.org/10.1177/0956797617719950.

Spoelstra, K. S., & van Doorn, A. (2019). Versneld kantelen naar Natuurinclusieve landbouw. *De Levende Natuur, 120*, 4.

Standing, G. (2019). *Plunder of the commons: A manifesto for sharing public wealth*. Penguin UK.

Stirling, A. (2006). Precaution, foresight and sustainability. Reflection and reflexivity in the governance of science and technology. In *Reflexive governance for sustainable development* (pp. 225–272). Cheltenham: Elgar.

van der Steen, M., Peeters, R., & van Twist, M. J. W. (2011). *De boom en het rizoom: overheidssturing in een netwerksamenleving, essay*. Den Haag: Ministerie van Volkshuisvesting, Ruimtelijke Ordening en Milieubeheer (VROM).

Steffen, W., et al. (2015). Planetary boundaries: Guiding human development on a changing planet. *Science, 347*(6223). https://doi.org/10.1126/science.1259855.

Sterman, J. D. (2000). *Business dynamics: Systems thinking and modeling for a complex world*. Boston: Irwin McGraw-Hill.

Stern, N. (2013, September). The structure of economic modeling of the potential impacts of climate change: Grafting gross underestimation of risk onto already narrow science models. *Journal of Economic Literature, 51*(3), 838–859.

Stiglitz, J. E. (2012). *The price of inequality: How today's divided society endangers our future*. New York: W.W. Norton & Company. ISBN: 9780393088694.

Stiglitz, J. E. (2015). *The great divide: Unequal societies and what we can do about them*. New York: W.W. Norton & Company. ISBN: 9780393248579.

Stiglitz, J. E. (2019a). *Measuring what counts: The global movement for well-being*. The New Press.

Stiglitz, J. E. (2019b). *People, power, and profits: Progressive capitalism for an age of discontent*. Penguin UK.

Stöckli, S., Dorn, M., & Liechti, S. (2018). Normative prompts reduce consumer food waste in restaurants. *Waste Management, 77*, 532–536. https://doi.org/10.1016/j.wasman.2018.04.047.

Stone, C. D. (1996). *Should trees have standing? And other essays on law, morals, and the environment* (Rev'd 2010 ed.). UK: Oxford University Press. ISBN: 978-0379213812.

Strengers, Y., & Maller, C. (Eds.). (2014). *Social practices, intervention and sustainability: Beyond behaviour change*. Routledge.

Stringer, L. C., Dougill, A. J., Fraser, E., Hubacek, K., Prell, C., & Reed, M. S. (2006). Unpacking 'participation' in the adaptive management of social ecological systems: A critical review. *Ecology and Society, 11*(2), 39.

Susskind, L., & Cruikshank, J. (2006). Anticipate the problems of following through. In *Breaking Robert's rule: The new way to run meetings, build consensus, and get results* (pp. 130–132). New York, NY: Oxford University Press.

Susskind, L., & Field, P. (1996). *Dealing with an angry public: The mutual gains approach to resolving disputes*. Simon and Schuster.

Sussman, L., & Bassarab, K. (2017). *Food policy council report 2016*. Baltimore, MD: Johns Hopkins University Center for a Livable Future.

Swyngedouw, E. (2009). The political economy and political ecology of the hydro-social cycle. *Journal of Contemporary Water Research & Education, 142*(1), 56–60.

Termeer, C. J. A. M. (2019). *Expertpaper over het bewerkstelligen van een transitie naar kringlooplandbouw*. Wageningen University & Research.
Termeer, C. J. A. M., & Dewulf, A. (2017, December). *Mogelijkheden van de 'smallwins' aanpak voor de transitie opgaven van het Ministerie van Infrastructuur en Waterstaat*. Leerstoelgroep Bestuurskunde, Wageningen University & Research.
Termeer, C. J., Dewulf, A., & Van Lieshout, M. (2010). Disentangling scale approaches in governance research: Comparing monocentric, multilevel, and adaptive governance. *Ecology and Society, 15*(4).
Termeer, C. J. A. M., Stuiver, M., Gerritsen, A., & Huntjens, P. (2013). Integrating self-governance in heavily regulated policy fields: Insights from a Dutch farmers' cooperative. *Journal of Environmental Policy & Planning, 15*(2), 285–302. https://doi.org/10.1080/1523908X.2013.778670.
The Economist. (2011). *Oh, Mr Porter*. The Economist. March 10, 2011. https://www.economist.com/business/2011/03/10/oh-mr-porter.
Theobald, E. J., Ettinger, A. K., Burgess, H. K., DeBey, L. B., Schmidt, N. R., Froehlich, H. E., et al. (2015). Global change and local solutions: Tapping the unrealized potential of citizen science for biodiversity research. *Biological Conservation, 181*, 236–244.
Therborn, G., & Aboim, S. (2014). The killing fields of inequality. *Análise Social, 212*, 729–735.
Thomas, J. M., Ursell, A., Robinson, E. L., Aveyard, P., Jebb, S. A., Herman, C. P., & Higgs, S. (2017). Using a descriptive social norm to increase vegetable selection in workplace restaurant settings. *Health Psychology, 36*(11), 1026–1033. https://doi.org/10.1037/hea0000478.
Tilly, C. (1998). *Durable inequality*. University of California Press.
Tirole, J. (2017, October). *Economics for the common good* (1st ed.). Princeton University Press, 576 p.
Tittonell, P., & Giller, K. E. (2012). When yield gaps are poverty traps: The paradigm of ecological intensification in African smallholder agriculture. *Field Crops Research, 143*, 76–90. Retrieved from https://www.klv.nl/media/uploads/skov_13_06_2013__litt._ken_giller_2__pdf.pdf.
Tompkins, E. L., & Adger, W. N. (2004). Does adaptive management of natural resources enhance resilience to climate change? *Ecology and Society, 9*(2).
Tonkiss, F. (2017). Urban economies and social inequalities. In *The SAGE handbook of the 21st century city* (p. 187). London: Sage.
Trainer, T. (2011). The radical implications of a zero growth economy. *Real-World Economics Review, 57*(1), 71–82.
Tromp, C. (2018, May). *Wicked philosophy: Philosophy of science and vision development for complex problems* (1st ed.). Amsterdam University Press, 240 p.
Turns, A. (2019, October 15). How citizen scientists, technology and people power can help clean up our seas. *Evening Standard*. Retrieved October 20, 2019, from https://www.standard.co.uk/futurelondon/theplasticfreeproject/how-citizen-scientists-technology-and-people-power-can-help-clean-up-our-seasa4261831.html
Tyler, J. A., & Swartz, A. L. (2012). Storytelling and transformative learning. In *The handbook of transformative learning: Theory, research, and practice* (pp. 455–470). San Francisco, CA: John Wiley & Sons.
Ulucak, R., & Kassouri, Y. (2020). An assessment of the environmental sustainability corridor: Investigating the non-linear effects of environmental taxation on CO_2 emissions. *Sustainable Development, 28*(4).
UN Brundtland Report. (1987). *Our common future*. Oxford: Oxford University Press.
UN Department of Economic and Social Affairs Population Division. (2015). Retrieved from https://esa.un.org/unpd/wup/Publications/Files/WUP2014-Report.pdf.
UN Secretary General. (2014, April 11). *Secretary-General's remarks at climate leaders summit*. Press release.
UN Secretary General. (2018, March 29). *Climate chaos to continue in 2018, UN chief warns; Will the world rise to challenge?* Press release.

UNEP. (2011). *Towards a green economy: Pathways to sustainable development and poverty eradication*. Nairobi, Kenya: UNEP.
UNESCAP. (2009). *What is good governance*. United Nations Economic and Social Commission for Asia and the Pacific. Retrieved March 16, 2020, from https://www.unescap.org/resources/what-good-governance
UNFPA. (2019). *State of the world population 2019. Unfinished business: The pursuit of rights and choices for all*. UNFPA Division of Communications and Strategic Partnerships.
UN-Habitat. (2020). *Future cities: New economy and shared city prosperity driven by technological innovations*. United Nations Human Settlements Programme (UN-Habitat).
United Nations. (2018). *World urbanization prospects 2018 revision*. New York: The Population Division of the Department of Economic and Social Affairs of the United Nations.
United Nations Development Programme. (2011). *Human development report 2011: Sustainability and equity: A better future for all*. New York: UNDP. Retrieved from http://hdr.undp.org/sites/default/files/reports/271/hdr_2011_en_complete.pdf.
UN-Water. (2013). *What is water security?* Infographic by UN-Water Publications. Retrieved June 17, 2020, from https://www.unwater.org/publications/water-security-infographic/.
Usher, R., & Bryant, I. (1989). *Adult education as theory, practice and research*. London: Routledge.
Van Buuren, A., & Edelenbos, J. (2004). Conflicting knowledge – Why is joint knowledge production such a problem? *Science and Public Policy, 31*(4), 289–299.
Van Buuren, A., Ellen, G. J., & Warner, J. F. (2016). Path-dependency and policy learning in the Dutch delta: Toward more resilient flood risk management in the Netherlands? *Ecology and Society, 21*(4).
Van den Bergh, J. C. (2001). Ecological economics: Themes, approaches, and differences with environmental economics. *Regional Environmental Change, 2*(1), 13–23.
Van den Bergh, J. C. (2017). A third option for climate policy within potential limits to growth. *Nature Climate Change, 7*(2), 107–112.
Van Diep, N., Khanh, N. H., Son, N. M., Van Hanh, N., & Huntjens, P. (2007). Integrated water resource management in the Red River Basin—Problems and cooperation opportunity. In *Proceedings of the CAIWA international conference on adaptive and integrated water management* (pp. 2–10). Basel, Switzerland.
Van Doorn, A., Melman, D., Westerink, J., Polman, N., Vogelzang, T., & Korevaar, H. (2016). *Natuurinclusieve landbouw, food-for-thought*. Wageningen: Wageningen University and Research.
Van Mierlo, B., Arkesteijn, M., & Leeuwis, C. (2010). Enhancing the reflexivity of system innovation projects with system analyses. *American Journal of Evaluation, 31*(2), 143–161.
Van Mierlo, B. C., Regeer, B., van Amstel, M., Arkesteijn, M. C. M., Beekman, V., Bunders, J. F. G., de Cock Buning, T., Elzen, B., Hoes, A. C., & Leeuwis, C. (2010). *Reflexive monitoring in action. A guide for monitoring system innovation projects*. Wageningen/Amsterdam: Communication and Innovation Studies, WUR/Athena Institute, VU.
Van Vliet, M. T., Vögele, S., & Rübbelke, D. (2013). Water constraints on European power supply under climate change: Impacts on electricity prices. *Environmental Research Letters, 8*(3), 035010.
Van Vuuren, D. P., Stehfest, E., Gernaat, D. E., Doelman, J. C., Van den Berg, M., Harmsen, M., et al. (2017). Energy, land-use and greenhouse gas emissions trajectories under a green growth paradigm. *Global Environmental Change, 42*, 237–250.
Varoufakis, Y. (2019, March). *Talking to my daughter about the economy: A brief history of capitalism*. Vintage Publishing, 224 p.
Velten, S., Leventon, J., Jager, N., & Newig, J. (2015). What is sustainable agriculture? A systematic review. *Sustainability, 7*(6), 7833–7865.
Velter, M., Bitzer, V., Bocken, N. M. P., & Kemp, R. (2020). Sustainable business model innovation: The role of boundary work for multi-stakeholder alignment. *Journal of Cleaner Production, 247*. https://doi.org/10.1016/j.jclepro.2019.119497.

Veneri, P., & Murtin, F. (2019). Where are the highest living standards? Measuring well-being and inclusiveness in OECD regions. *Regional Studies, 53*(5), 657–666.
Verhees, G. (2013). *Publiek-private samenwerking: adaptieve planning in theorie en praktijk.* Proefschrift Rijksuniversiteit Groningen, Faculteit: Ruimtelijke Wetenschappen.
Vermaak, H. (2009). *Plezier beleven aan taaie vraagstukken.* Kluwer.
Visser, W. (2017, September 30). Integrated value: What it is, what it's not and why it's important. *HuffPost.*
Visser, W., & Kymal, C. (2015). Integrated value creation (IVC): Beyond corporate social responsibility (CSR) and creating shared value (CSV). *Journal of International Business Ethics, 8*(1), 29–43.
Visseren-Hamakers, I. J. (2020). The 18th sustainable development goal. *Earth System Governance.* https://doi.org/10.1016/j.esg.2020.100047.
Vissers, G. A. N., Heyne, G. A. W. M., & Peters, V. A. M. (1995). Spelsimulatie in bestuurskundig onderzoek. *Bestuurskunde, 4*(4), 178–187.
Von Borgstede, C., Johansson, L.-O., & Nilsson, A. (2013). Social dilemmas: Motivational, individual, and structural aspects influencing cooperation. In L. E. Steg, A. E. Van Den Berg, & J. I. De Groot (Eds.), *Environmental psychology: An introduction* (pp. 175–184). BPS Blackwell.
Voorberg, W. H., Bekkers, V. J., & Tummers, L. G. (2015). A systematic review of co-creation and co-production: Embarking on the social innovation journey. *Public Management Review, 17*(9), 1333–1357.
Voß, J. P., & Bornemann, B. (2011). The politics of reflexive governance: Challenges for designing adaptive management and transition management. *Ecology and Society, 16*(2).
Voß, J. P., & Kemp, R. (2015). *Sustainability and reflexive governance: Introduction.* Technische Universität Berlin.
Wahl, D. (2016). *Designing regenerative cultures.* Triarchy Press.
Walker, B., Holling, C. S., Carpenter, S., & Kinzig, A. (2004). Resilience, adaptability and transformability in social–ecological systems. *Ecology and Society, 9*(2), 5. [Online]. Retrieved from http://www.ecologyandsociety.org/vol9/iss2/art5.
Wallace, R. G., Bergmann, L., Kock, R., Gilbert, M., Hogerwerf, L., Wallace, R., & Holmberg, M. (2015). The dawn of structural one health: A new science tracking disease emergence along circuits of capital. *Social Science & Medicine, 129,* 68–77.
Walters, C. (1986). *Adaptive management of renewable resources.* New York, NY: Macmillan and Co.
Weale, A. (2011). New modes of governance, political accountability and public reason 1. *Government and Opposition, 46*(1), 58–80.
Weale, A. (2004). Contractarian theory, deliberative democracy and general agreement. In K. Dowding, R. Goodin, C. Pateman, & B. Barry (Eds.), *Justice and democracy: Essays for Brian Barry* (pp. 79–96). Cambridge: Cambridge University Press.
Webster, K. (2013). What might we say about a circular economy? Some temptations to avoid if possible. *World Futures, 69*(7–8), 542–554.
Webster, K. (2014). *The circular economy: A wealth of flows.* Cowes, UK: Ellen MacArthur Foundation.
Weible, C. M., Sabatier, P. A., & McQueen, K. (2009). Themes and variations: Taking stock of the advocacy coalition framework. *The Policy Studies Journal, 37*(1).
Weick, K. E. (1984). Small wins: Redefining the scale of social problems. *American Psychologist, 39*(1), 40–49.
Wendt, A. (1987). The agent-structure problem in international relations theory. *International Organization, 41*(3), 335–350.
Wertheim-Heck, S., Raneri, J. E., & Oosterveer, P. (2019). Food safety and nutrition for low-income urbanites: Exploring a social justice dilemma in consumption policy. *Environment and Urbanization, 31*(2), 397–420.

Westerink, J., de Boer, T. A., Pleijte, M., & Schrijver, R. A. M. (2019). *Kan een goede boer natuurinclusief zijn?: De rol van culturele normen in een beweging richting natuurinclusieve landbouw* (2352–2739). Wageningen. Retrieved from https://edepot.wur.nl/508108.

WFP. (2017). *At the root of the exodus: Food security, conflict and international migration.* Rome: WFP.

White, T. (2020). Direct producer-consumer transactions: Community supported agriculture and its offshoots. In *The handbook of diverse economies.* Edward Elgar Publishing.

White, M. P., Alcock, I., Grellier, J., Wheeler, B. W., Hartig, T., Warber, S. L., et al. (2019). Spending at least 120 minutes a week in nature is associated with good health and wellbeing. *Scientific Reports, 9*(1), 1–11.

WHO. (2015). *Sheet No. 310* [Website]. Geneva: World Health Organization. Retrieved December 4, 2015, from http://www.who.int/me-diacentre/factsheets/fs310/en/index2.html.

WHO. (2016). *Global report on urban health: Equitable healthier cities for sustainable development.* World Health Organization.

Wielinga, E., & Robijn, S. (2018). *Netwerken met energie: Gereedschap voor co-creatie* (1st ed.). Scriptum. EAN: 9789463191159.

Wijnen, M., Augeard, B., Hiller, B., Ward, C., & Huntjens, P. (2012). *Managing the invisible: Understanding and improving groundwater governance. Water papers.* Washington, DC: World Bank. Retrieved from https://openknowledge.worldbank.org/handle/10986/17228.

Willems, M., & Roelofs, E. M. G. (2009, September 10–12). *Learning history as an evaluation method for the policy formulation of the Dutch Societal Innovation Agenda on Energy.* Paper for the panel session 'New conceptual and normative approaches for the evaluation of public policy/provisions' of the European Consortium of Policy Research Conference in Potsdam, Germany.

Willett, W., Rockström, J., Loken, B., Springmann, M., Lang, T., Vermeulen, S., et al. (2019). Food in the Anthropocene: The EAT–Lancet Commission on healthy diets from sustainable food systems. *The Lancet, 393*(10170), 447–492.

Williams, R., & Hayes, J. (2013). Literature review: Seminal papers on shared value. *EPS-PEAKS.*

Williams, L. D. A., & Woodson, T. S. (2012). The future of innovation studies in less economically developed countries. *Minerva, 50*(2), 221–237. https://doi.org/10.1007/s11024-012-9200-z. JSTOR 43548641.

Witkamp, M. J., Raven, R. P., & Royakkers, L. M. (2011). Strategic niche management of social innovations: The case of social entrepreneurship. *Technology Analysis & Strategic Management, 23*(6), 667–681.

Wittmayer, J. M., Avelino, F., van Steenbergen, F., & Loorbach, D. (2017). Actor roles in transition: Insights from sociological perspectives. *Environmental Innovation and Societal Transitions, 24,* 45–56.

Wittmayer, J. M., Backhaus, J., Avelino, F., Pel, B., Strasser, T., Kunze, I., & Zuijderwijk, L. (2019). Narratives of change: How social innovation initiatives construct societal transformation. *Futures, 112,* 102433.

WMO (2019) World Meteorological Organization (WMO); United Nations Environment Programme; Intergovernmental Panel on Climate Change; Global Framework for Climate Services (GFCS). *United In Science: High-level synthesis report of latest climate science information convened by the Science Advisory Group of the UN Climate Action Summit 2019.*

World Bank. (2010). *Development and climate change. World development report 2010.* Washington, DC: World Bank.

World Economic Forum. (2020, January 15). *The global risks report 2020.* World Economic Forum, in partnership with Marsh & McLennan and Zurich Insurance Group.

WRR. (2014). Rapport 93. Naar een voedselbeleid. https://www.wrr.nl/publicaties/rapporten/2014/10/02/naar-eenvoedselbeleid.

Xepapadeas, A. (2008). Ecological economics. In *The new Palgrave dictionary of economics* (2nd ed.). Palgrave Macmillan.

Yasuda, Y., Aich, D., Hill, D., Huntjens, P., & Swain, A. (2017a). *Transboundary water cooperation over the Brahmaputra River: Legal political economy analysis of current and future potential cooperation.* Hague, the Netherlands: The Hague Institute for Global Justice.

Yasuda, Y., Schillinger, J., Huntjens, P., Alofs, C., & De Man, R. (2017b). *Transboundary water cooperation over the lower part of the Jordan River Basin: Legal political economy analysis of current and future potential cooperation.* The Hague Institute for Global Justice. Retrieved from https://www.siwi.org/publications/transboundary-water-cooperation-lower-part-jordan-river-basin/.

Yasuda, Y., Hill, D., Aich, D., Huntjens, P., & Swain, A. (2018). Multi-track water diplomacy: Current and potential future cooperation over the Brahmaputra River Basin. *Water International, 43*(5), 642–664. https://doi.org/10.1080/02508060.2018.1503446.

Yasuda, Y., Hill, D., Aich, D., Huntjens, P., & Swain, A. (2020). Multi-track water diplomacy: Current and potential future cooperation over the Brahmaputra River Basin. In *A river flows through it: A comparative study of transboundary water disputes and cooperation in Asia.* New York: Routledge, 2020/12/18, 159 pages.

Young, O. R. (2002). *The institutional dimensions of environmental change: Fit, interplay, and scale (Global environmental accords: Strategies for sustainability).* Cumberland: MIT Press.

Zartman, W., & Rubin, J. (2000). Symmetry and asymmetry in negotiation. In W. Zartman & J. Rubin (Eds.), *Power and negotiation* (pp. 271–294). Ann Arbor, MI: University of Michigan Press.

Zevenbergen, C., Rijke, J., Van Herk, S., & Bloemen, P. J. T. M. (2015). Room for the river: A stepping stone in adaptive delta management. *International Journal of Water Governance, 3*(1), 121–140.

Ziegler, R. (2013). Real social contracts for sustainability? Philosophical and political implications of social agreement in circumstances of poverty and degraded ecosystems. *Public Reason, 5*(2), 3–20.

Zietsma, C., & Lawrence, T. B. (2010). Institutional work in the transformation of an organizational field: The interplay of boundary work and practice work. *Administrative Science Quarterly, 55*(2), 189–221.

Zoeteman, K., & Tavenier, J. (2012). A short history of sustainable development (Chapter 2). In K. Zoeteman (Ed.), *Sustainable development drivers* (pp. 14–54). Cheltenham: Edward Elgar.

Zuidhof, P. W. (2014). De Wonderlijke Geschiedenis Van het Neoliberalisme: Alles Wat Je Ooit Wilde Weten over het Neoliberalisme Maar Nooit Durfde te Vragen. *Groniek: Historisch Tijdschrift, 199*, 193–210.

The manufacturer's authorised representative in the EU is Springer Nature Customer Service Centre GmbH, Europaplatz 3, 69115 Heidelberg, Germany. If you have any concerns regarding our products, please contact ProductSafety@springernature.com

Printed and bound by CPI Group (UK) Ltd, Croydon, CR0 4YY

23/03/2026

02076661-0005